United States–China Technology Transfer

United States–China Technology Transfer

Otto Schnepp
Mary Ann Von Glinow
Arvind Bhambri

UNIVERSITY OF SOUTHERN CALIFORNIA

Prentice Hall, Englewood Cliffs, New Jersey 07632

Library of Congress Cataloging-in-Publication Data

SCHNEPP, OTTO.
 United States–China technology transfer / Otto Schnepp, Mary Ann
Von Glinow, Arvind Bhambri.
  p. cm.

 Includes bibliographical references.
 ISBN 0-13-949975-X
 1. Technology transfer—United States. 2. Technology transfer-
China. I. Von Glinow, Mary Ann Young,   II. Bhambri,
Arvind. III. Title.
T174.3.S33 1990
338.97306—dc20   89–49010
                   CIP

Editorial/production supervision
 and interior design: **Lorraine Antine**
Cover design: **Photo Plus Art**
Manufacturing buyer: **Peter Havens**

 © 1990 by Prentice-Hall, Inc.
A Division of Simon & Schuster
Englewood Cliffs, New Jersey 07632

Printed in the United States of America
10 9 8 7 6 5 4 3 2 1

ISBN 0-13-949975-X

Prentice-Hall International (UK) Limited, *London*
Prentice-Hall of Australia Pty. Limited, *Sydney*
Prentice-Hall Canada Inc., *Toronto*
Prentice-Hall Hispanoamericana, S.A., *Mexico City*
Prentice-Hall of India Private Limited, *New Delhi*
Prentice-Hall of Japan, Inc., *Tokyo*
Simon & Schuster Asia Pte. Ltd., *Singapore*
Editora Prentice-Hall do Brasil, Ltda., *Rio de Janeiro*

Contents

PART II   CASE STUDIES

Case 1

Case 2

Case 3

Case 4

Case 5

Case 6

Case 7

Case 8

PART III   COMPARATIVE EVALUATION

Chapter 3

Chapter 4

Implications for Management   *239*

Appendix

Outline of Technology Transfer Case Studies   *255*

References   *257*

Preface

United States–China Technology Transfer is a book about the issues that arise at different stages during the transfer of technology between different cultures. Using U.S.–China technology transfer as the focus, it raises implications that are equally relevant to managing the technology transfer process with East European countries and less developed countries (LDC's). Problems of language, cultural differences, infrastructural constraints, different standards, and politicization of the technology transfer agreement complicate the transfer process in each of these settings. The methods by which companies attempt to overcome these hurdles (or resolve these problems) make up the essence of this book.

The book describes the strategies and tactics used by four American firms, as well as the four Chinese enterprises that received the technology, in the transfer of design and manufacturing know-how. It charts their successes as well as their difficulties in bridging the cultural, political, and economic gaps that have stalled many U.S. firms anticipating entry into China. What differentiates this book from the numerous other books on international technology transfer is its attempt to tell both sides of the technology transfer story—the Chinese side as well as the U.S. side. Further setting this book apart from similar texts is its description of technology transfer as a series of four discrete stages: (1) development; (2) negotiations; (3) start-up (implementation); and (4) management of the ongoing process.[1] By focusing deliberately on each stage, we analyze the aspects characteristic of each stage and highlight those that pertain to more than one stage.

Our findings are based on exhaustive field research conducted in the U.S. and in China between 1984 and 1987. Since the political events of May/June 1989 and their aftermath, the political situation in China has changed dramatically and will certainly affect the country's short term trade relations with the United States. We discuss these events and their likely impact. More importantly, our book focuses on the microdynamics of implementing technology transfer, the nuts and bolts of managing the transfer of expertise through people, documentation and equipment, when confronted with cultural and other hurdles. This is the essential contribution of our work and we believe that it remains relevant in the face of recent events in China.

The book has three major parts. Part I deals with the background and framing of the key questions underlying this research. We examine the current trends in technology transfer research as well as the environment for technology acquisition in China. China has undergone one of the most dramatic turnarounds, from the Mao era, which ended in 1976, to the "open door" policy, which led to economic and enterprise reform of the 1980's and, recently, a possible reversal and tightening of these policies. We examine these changes and their implications for technology transfer to China.

Part II forms the core of our field research. Here we present eight case studies, with their rich detail on how technology transfer has actually occurred. We describe how the pioneer firms—Foxboro, Combustion Engineering, Cummins Engine Company, and Westinghouse Electric Corporation—entered China in their respective markets and transferred design and manufacturing technology. We also describe their strategic choices and methods of implementation. In addition, we describe the priorities, concerns, and experiences of the Chinese enterprises receiving the technology and of their leading government agencies.

Part III, the final section, contains a comparative discussion of the different strategies and the relative effectiveness of these U.S. firms during each stage of the technology transfer process. We identify the key success criteria employed and assess how each firm fared on each criterion. One of the goals in this part is to offer insights on "how to" do business in China to managers who are either currently involved in the China market or who are anticipating entry. Our insights focus on the concern for the effective implementation of technology transfer across any two disparate cultures.

We have made an effort to present a large volume of information, both practical and research-focused, in readable form. In making our choices of included material, we tried to address the needs of managers, instructors, and researchers. We have developed a four-stage model of the technology transfer process, which we offer as a first attempt at developing a taxonomy of the com-

---

1. See Appendix for a more detailed outline of these stages.

plex processes inherent in U.S.–China technology transfer. This framework permits a focused view of a complex process at a particular point in time. Further research should specifically address issues of technology transfer within the context of the four stages.

We are indebted to the U.S. Department of Education for its help in funding this research effort. We are also grateful to the executives from Foxboro, Combustion Engineering, Cummins Engine Company, and Westinghouse Electric Corporation for their willingness to cooperate with us in preparing cases about their experience in China. We are particularly indebted to the officials from various national agencies in the People's Republic of China—the Ministry of Machine Building Industries, the State Science and Technology Commission, and the National Research Center for Science and Technology for Development—and from various local institutes, agencies, and Chinese enterprises in Shanghai, Beijing, Harbin, and Chongqing. Without their formal support and willingness to sponsor us, this research would not not have been possible. We especially thank Dr. Bai Yiyan for all his efforts on our behalf. We thank Mary B. Teagarden and Bryan Williamson for their able editorial and research assistance. Finally, we thank our spouses—Judith Stransky, Steve Kerr, and Shobita Misra—for bearing with us during this prolonged project.

ACKNOWLEDGMENTS

The authors acknowledge the valuable assistance of many individuals who contributed so generously to our work. Without their help, our research would not have been possible.

Foxboro Company

Gerald Gleason, Vice President

Ernest J. DeBellis, Manager, International Sales Planning

Lawrence C. Martin, Manufacturing Division Controller

Frederick T. Morse, Corporate Manufacturing Consultant

Edward McIntyre, Head, Legal Department

Thomas A. Stuhlfire, Manager, Special Services

Donald Sorterup, former General Manager, Shanghai Foxboro Company Limited (SFCL)

Charles Rathke, Chief Engineer, Shanghai Foxobro Company Limited (SFCL)

Westinghouse Electric Company

Robert Murphy, Director, China Program, Steam Turbine Generator Division

Clovis F. Obermesser, Vice President, and President, Asia-Pacific

George J. Dubrasky, Management Representative, China, Power Generation Commercial Division

Robert T. Winston, Director, International Trade Policy

Richard C. Gaskins, Director, China Programs

John C. Denman Jr., Chief Counsel, Asia-Pacific

C. George Butterfield, Director, China Operations, Power Generation, Commercial Division

Carl I. Hammer Jr., Associate Director, Business Environment Assessment, Corporate Planning

Francis Kao, Physicist, Steam Turbine Generator Division

Robert W. Gray, Manager, Project Finance, Treasury Department

Gary D. Albrecht, Director, International Planning, Steam Turbine Generator Division

John G. Salvati, Engineering Section Manager, Components Technology, Technology and Quality Division

Charles T. Ng, Representative, Beijing Office

Cummins Engine Company

Dr. Andrew Chu, General Manager, China Business Group

Thomas W. Head, Vice President and General Manager, Affiliated Enterprises

Charles B. Byers, Vice President, International Business Development

Dennis Kelly, formerly Director of Operations, China Business Group

Dennis Piper, Manager, Engineering Liaison, China Business Group

Combustion Engineering Inc.

Joseph F. Condon, Vice President, International

Frank W. Marshall, Vice President, Licensing, Joint Ventures and Associated Companies, International Division

Alexander Sivas, Vice President, Business Development and Sales, International Division, Power Systems

Robert D. Haun, Project Manager, International Division

Samuel S. Blackburn, Jr., Director, Licensing Engineering, Power Systems

Anthony J. Czapracki, International Division-Licensing, Power Systems

Josef W. Cumpelik, Vice President, Regional Officer—Asia/Pacific

James L. Van Fleet, Director, China, Power Systems

Barbara B. Van Fleet, Office Manager, China Office, Power Systems

Lawrence J. Froot, Regional Manager, Business Development, Power Systems

Steyr-Daimler-Puch, AG

Schaefl, Franz, Beijing Representative, Export Division

Chinese Organizations

Wu Wufeng, Commissioner, Secretary General, State Science and Technology Commission

Xu Youyu, Executive Deputy Director, China National Research Center for Science and Technology for Development

Xu Zhaoxiang, Deputy Director, China National Research Center for Science and Technology for Development

Dr. Bai Yiyan, Division Chief, Division of Technology, China National Research Center for Science and Technology for Development

Song Juzhi, Deputy Director, Science and Technology Bureau, Ministry of Machine Building Industry, and Director, Chinese Mechanical Engineering Society

Zhang Songlin, Senior Engineer, Instrumentation Industry Bureau, Ministry of Machine Building Industry

Li Peizhang, Deputy Chief Engineer, Electrical Equipment Bureau, Ministry of Machine Building Industry

He Yili, Senior Engineer, Section Chief, Foreign Economic Relations and Technical Cooperation Department, Electrical Equipment Bureau, Ministry of Machine Building Industry

Liu Xianzeng, Senior Engineer, China National Automotive Industry Corporation (formerly called Automotive Bureau), Ministry of Machine Building Industry

Hu Liang, Senior Advisor, China National Automotive Bureau, and Chairman, China Automotive Engineers Association, Ministry of Machine Building Industry

Wu Jiatai, Vice President, China National Technical Import Corporation, Technical Trading Consultant Company

Wu Yiquan, Chief Engineer, Harbin Boiler Works

Shen Jinru, Vice Chief Engineer, Harbin Boiler Works

Chang Fengshu, Senior Engineer, Harbin Electrical Machinery Works

Chu Pinchang, Chief Engineer, Harbin Power System Engineering and Research Institute

Xu Damao, Chief Engineer, Harbin Turbine Works

Wang Yintong, Shanghai Academy of Social Sciences, Economic, Legal and Social Consultancy Center

Yan Kuang'guo, Director, Shanghai Academy of Social Sciences, Economic, Legal and Social Consultancy Center

Wang Dawei, Chief Accountant, Shanghai Foxboro Company Limited (SFCL)

Chu Ruifeng, Deputy General Manager, Imports and Exports Department, Shanghai Instrumentation Corporation

Wang Minghui, Senior Engineer, Imports and Exports Department, Shanghai Instrumentation Corporation

Wang Zhenfeng, Deputy General Manager, Shanghai Instrumentation and Electronics Import and Export Corporation

Wei Jianxin, Chief Engineer, Shanghai Analytical Instrument Factory

Ke Jiaxiang, Deputy Chief, Imports and Exports

Cheng Shengmin, Deputy Chief Engineer, Shanghai Boiler Works

Wang Canghai, Manager, International, Shanghai Boiler Works

Huang Deshu, Manager, T-G Design Engineering, Shanghai Electric Machinery Manufacturing Works

Yao Songyao, Chief Engineer, Shanghai Electric Power Equipment Corporation

Zhou Xisheng, Vice Director and Chief Engineer, Shanghai Power Equipment Research Institute

Tao Dingwen, Vice Chief Engineer, Shanghai Turbine Works

Chai Jingdong, Executive Deputy Plant Manager, Chongqing Automotive Engine Plant

Yi Jizeng, Deputy Chief Engineer, Chongqing Automotive Engine Plant

United States–China
Technology
Transfer

CHAPTER 1

Introduction

SIGNIFICANCE OF U.S.-CHINA TECHNOLOGY TRANSFER

One of the major priorities of top U.S. executives in the 1980s has been to establish a foothold in the newly opened China market. They have viewed China as the last major undeveloped market in the world, especially for industrial equipment. And as large industrial markets have matured in the Western world, with resulting saturation and declining growth, the prospect of entering the vast, untapped China market in the early stages of its development cycle has generated immediate interest. The interest of Western executives has been reciprocated by Chinese policy makers and officials. After decades of isolation, in 1977 and 1978 Chinese leaders opened up to the outside world and set ambitious goals for the country's modernization and technological development. In 1981, Premier Zhao Ziyang announced the aim to quadruple the gross annual value of industrial and agricultural production by the year 2000. He also advocated qualitative change of the means of production through technological innovation and improved management. Existing plants in key sectors were to be modernized with imported technology. In addition, broader reforms in enterprise autonomy and decision-making power were to be implemented.

The change in government policy has been reflected in China's trade statistics. In 1970, China's total foreign trade was U.S. $4.59 billion; by 1978 it had increased to U.S. $20.64 billion, and it reached U.S. $102.8 billion in

1

1988.[1] Moreover, the capitalist countries of the industrialized world have been China's major trading partners. For the period 1979 to 1986, Japan had 24.9 percent of China's trade; EEC countries had 22.6 percent; Hong Kong, 21.5 percent; and the United States, 11.8 percent.

The Chinese government has also continued to stress its aim of technology absorption through trade and technology agreements in the form of licensing and equity joint ventures. In 1986, excluding joint ventures, the Ministry of Foreign Economic Relations and Trade (MOFERT) approved 744 technology transfer contracts, for a total value of $4.46 billion, an increase of 50.7 percent over the previous year's value. The U.S. share of technology imports in 1986 was $657 million, or 14.7 percent. When one considers foreign-invested enterprises in China, the United States is the second leading source of capital, after Hong Kong. Between 1979 and 1985, investment commitments by U.S. companies amounted to $2.1 billion; during 1986-87, an additional $1 billion was invested. By end-1985, the United States had 138 equity joint ventures and an additional 81 were contracted in 1986. Thus, these investment and trade statistics confirm a continuing effort by U.S. companies to enter and become established in the China market. Moreover, at the core of most long-term U.S.-China projects is the transfer of technology, which includes manufacturing, design, and management expertise.

In technology transfer, expertise and knowledge are transmitted among individuals and organizations. Software, as embodied in human expertise and in documentation, and hardware, such as materials, components, and equipment, are transferred. Effective transfer requires a consideration of multiple factors including the differences in culture, decision-making processes and styles, technical standards, manufacturing infrastructure, language, and management practices. Not surprisingly, U.S. executives engaged in technology transfer to China have experienced frequent frustrations and delays. Moreover, existing research and models of technology transfer have proven inadequate in helping with the unique idiosyncrasies of U.S.-China projects.

TECHNOLOGY TRANSFER—THE ACADEMIC VIEW

The concept of technology and technology transfer has been defined in many ways, measured in still different ways, and assessed against highly diverse criteria (Stewart and Nihei, 1987). Questions such as the following have dominated discussions of technology transfer: What do we mean by technology transfer? What are the processes whereby transfer of technology occurs? What are the roles of different institutions in technology transfer? What factors should be considered in the amount and type of technology to be transferred? What constraints exist upon the execution of technology transfer, and how can these be

---

[1]All dollar figures are in U.S. dollars throughout this text unless otherwise noted.

relaxed? Stewart and Nihei (1987, p. 1) puzzle through some of these questions and begin with a basic definition of technology: "Technology refers to new and better ways of achieving economic ends that contribute to economic development and growth."

The emphasis on the role of technology in the economic arena is important, and most theorists today also anchor technology transfer in the role of the economy, and progress (Contractor and Sagafi-Nejad, 1981; Gee, 1981; Marton, 1986; Pugel, 1978). Most contemporary discussions of technology transfer assume that technology and its advance contribute to change and that this change fuels economic growth through productivity increases (Gee, 1981). Thus, when we refer to technology transfer, we refer to the application of technology to a new use or user (Gee, 1981) for economic or productivity gains. Gruber and Marquis (1969) note further that new techniques must often be used where they have not been previously used. Technology transfer can thus be defined as a process by which expertise or knowledge related to some aspect of technology is passed from one user to another for the purpose of economic gain.

Technology transfer of the type described above has been referred to as *horizontal,* that is, from one organization to another. *Vertical* technology transfer occurs between different stages of a particular innovation process (Teece, 1976). We also distinguish between technology transfer and diffusion, in that like Stewart and Nihei (1987), we believe that technology transfer concentrates on the supply side, or the ability and willingness of a supplier to transfer, whereas diffusion says little about technology supply and concentrates on the demand characteristics (p. 2). Conceptually, however, both technology transfer and diffusion are closely related.

It is the purpose of this chapter to examine the existing conceptualizations of technology transfer and apply relevant points to U.S.-China technology transfer. This research backdrop is then used to inform our research questions, develop our research framework, and launch our case investigations on U.S.-China technology transfer. The remainder of this chapter analyzes the major trails of research on technology transfer, particularly as they apply to the U.S.-China context and details our research framework in preparation for our case studies in Part II.

CURRENT TRENDS IN THE RESEARCH LITERATURE

Research on technology transfer, particularly theoretical research, is still in a very early development stage. Jolly and Creighton (1975) point out that as of the 1970s, there were only a limited number of centers engaged in research on technology transfer. Bell and Hill (1978) went on to claim further that the field in general lacks acceptable taxonomic development. It has been noted, with some dismay, that similar problems have persisted into the 1980s (Godkin, 1988). The earliest documented research on technology transfer was by anthropological and sociological researchers (Rogers and Shoemaker, 1971) and economists during the 1950s and 1960s (Godkin, 1988). These researchers sought explanations for different economic growth rates, and rationales for the failures of U.S. technology in the "transplant" process abroad during that time period (Jeremy, 1981). Despite an abundance of innovation and diffusion-related research, Boyle (1986) concluded that the bulk of current research on technology transfer was arigorous, nonempirical, and largely anecdotal and normative. Descriptive studies were largely absent from the research literature. Thus, it is against this background of paucity of rigorous work that we attempt to classify the extant research on technology transfer.

Kedia and Bhagat (1988) note that three distinct conceptual trails of research exist on technology transfer. The first includes studies that examine the processes by which technology gets transferred, the types of technology that are likely to be received more successfully by users, and the on-going relationships of the two participants as dictated by early negotiations (Balasubramanyam, 1973; Marton, 1986; Mason, 1980). The second set of research studies focuses on absorptive capacities of the recipient enterprises, as well as technological developments of the host countries (Baranson, 1970; Driscoll and Wallender, 1981; Dunning, 1981; Grow, 1987, 1988). The third category, and clearly the dominant category, includes case-based research that examines the effectiveness of technology transfer based on a variety of factors including industry characteristics, technological sophistication of supplier and recipient organizations, and technology maturity (Baranson and Harrington, 1977; Behrman and Wallender, 1976; Evenson, 1976; Marton, 1986; Von Glinow, Schnepp, and Bhambri, in press).

More recently there have been attacks levied at the extant research literature on technology transfer for failure to include contextual or culture-based variations that account for success or failure (Kedia and Bhagat, 1988; Von Glinow and Teagarden, 1988).

Research on Processes by Which Technology Gets Transferred

This first category of research studies primarily focuses on technology transfer from a developed country, usually in the West, to a developing country, usually in one of the Third World nations (Kedia and Bhagat, 1988). The emphasis here is upon process, that is, the way in which product-embodied, process-embodied, or person-embodied technologies are transferred. Product-embodied refers to the transfer of a product, or specific hardware without the transfer of knowledge or skills required to manufacture or design the product. Process-embodied technology refers to the establishment of a process, for example, the automation of some technology or a turnkey plant, without necessarily going into the design or R&D of it. Finally, person-embodied refers to the knowledge or skills possessed by individuals. A major distinction made by Hall and Johnson (1970) is that the process of technology transfer is significantly different from the transfer of a commodity. Moreover, the relative success or effectiveness of a given transfer depends on the type of technology involved, that is, whether it involves products, processes, or people. It has been argued that technologies involving processes or people are much more difficult to transfer successfully than product-embodied technologies (Davidson, 1987; Kedia and Bhagat, 1988; Von Glinow, Schnepp, Bhambri, in press).

Godkin (1988) notes that there are a number of factors that have been determined empirically to influence the success of the technology transfer process: management support, appropriate legislation, brokering systems (or third-party individuals or groups to assist targeted user groups as a supplement to in-house staffs), and interaction among research universities, governments, and industries.

Technology is more likely to be received successfully by users, particularly when the user is a developing country, when the acquisition and absorption of industrial technology are isomorphic with the economic and business goals of the developing country and the respective recipient enterprise (Grow, 1988; Von Glinow, Schnepp, Bhambri, in press). Questions of contextual fit arise here, as well as in the second general category of research studies on absorptive capacities (Bradbury et al., 1978; Godkin, 1988). Finally, research pertaining to the quality of the ongoing relationship between supplier and recipient, as a function of the early negotiations process, has been discussed at length by Pye (1982).

Research on the Absorptive Capacities of Recipient Enterprises

Shifting the focus to the recipient enterprise we note that absorptive capacity is frequently a function of the perceived "fit" between the sender and the recipient (Grow, 1987, 1988). Since the emphasis here is upon the recipient's ability to absorb and utilize technology, it is useful to distinguish among different levels of technology. Dahlman and Westphal (1983) describe three degrees of capability: (1) operating a technology; (2) creating new productive capacity by investment; and (3) making innovations or changing and improving existing methods and products (Stewart and Nihei, 1987). These authors claim that each level requires different skills and supporting institutions, yet "only when all three capabilities have been transferred has the receiving nation acquired a permanent mastery of the technology. Operational capability can be learned on the job. Investment capability requires formal training; on-the-job training is protracted, and what is learned needs to be adapted, not just replicated. Innovative capability cannot be acquired from work experience alone" (pp. 3–4).

Stewart and Nihei (1987) further note that there are other conditions that affect technology transfer: general knowledge acquired through education; the propensity to adopt new techniques; and the general values, attitudes, and institutions that hinder or facilitate technology transfer. General knowledge is said to form the core of absorptive capacity in that it prepares people to acquire a specific technology required for economic growth. Research literature fails to distinguish between this concept of general knowledge as a prerequisite to absorption and the technology-specific knowledge necessary for absorption (Stewart and Nihei, 1987).

The ability or propensity to adopt new products appears to be correlated with general knowledge, but it is expressed by a country's willingness to accept change. Most certainly attitudes and values inherent in the religion or culture of a country can foster or inhibit change (Hofstede, 1980). Hofstede's work (1980) is perhaps the most frequently cited work focusing on country attitudes and values, and, while heavily criticized, it is relevant here insomuch as change or modification of such attitudes and values may be a precondition for accelerated technology absorption (Niehoff and Anderson, 1964; Novack and Lekachman, 1964; Stewart and Nihei, 1987).

Absorptive capacity clearly has many components and appears to be the most diffuse category of research on technology transfer. In the aggregate, absorptive capacity refers to a country's supply of professionals and workers who possess the general educational background for training. In addition, it includes the state of the legal, social, and economic infrastructure of the country that facilitates or hinders the assimilation of new expertise (Stewart and Nihei, 1987). This second factor in absorptive capacity is predicated on the first and establishes limits concerning how much and what type of technology can be acquired at acceptable risks and costs. It has been noted that

> No developing country can make adequate use of technology flows unless it has an appropriate receiving system composed of a scientific and technical infrastructure in the public, private, educational, and corporate sectors. It is of no value to develop a technical training system or R&D institutes when companies are not interested in using them (Fund for Multinational Management Education, 1978, vol. 2, p. 55).

Within the last five years, there has been a call for recognition of cultural or contextual variations across nations (Kedia and Bhagat, 1988). This has been previously implicitly alluded to in the research literature (Koizumi, 1982); however, it has rarely been empirically assessed or tested. It is widely assumed that the culture and the cultural context of a nation make a difference in the success of technology transfer to that nation; however, cultural variations tend to be difficult to assess with any degree of precision across nations (Hofstede, 1980). Kedia and Bhagat (1988) call for recognition of cultural variations across supplier and recipient firms, in that absorptive capacities may be contingent on cultural compatibility (Adler, 1986) between the partners.

Case-Based Research

Clearly the largest category of investigation into the success or failure of technology transfer is the case-based category. This category of research investigates the number of ways technology transfer may be accomplished on a case-by-case basis. The lens is broad; technology transfer may be accomplished through educational processes, turnkey operations, licensing agreements, joint ventures, trade shows, direct sales, and consulting. Each method is said to have its advantages and disadvantages, and no single method appears to be effective across all case situations (Fulk, Rogers, and Von Glinow, 1988).

Case studies involving technology transfer from one country to another are myriad; those most frequently cited are those from developed countries to less-developed countries (LDCs), for example, from the United States, Japan, or an EEC country to an LDC. There have been a number of observed differences among these three groupings of industrialized countries in their investment postures in LDCs (Campbell, 1986). Host country policies concerning technology transfers tend to be more important than the characteristics of the products being transferred. For example, national governments can affect the amount, type, and duration of technology transfers by restricting ownership of foreign firms (Stewart and Nihei, 1987).

Most case studies document that firms prefer to produce in and export from the home country, with the transfer of technology usually limited to product-embodied technology and to training and consulting to support customer service. Moving up the scale, evidence suggests that foreign investment by technology suppliers is generally the second choice. U.S. firms investing abroad have strong preferences for wholly owned subsidiaries, followed by majority ownership, and finally minority ownership. Non-equity-based licensing tends to be considerably further down the list (Stewart and Nihei, 1987).

Apart from the structural forms through which technology gets transferred, the willingness of a supplier firm to transfer technology largely depends upon its expected return, as well as its opportunity costs. Transfer is somewhat easier late in the technological product life cycle than when the product is new (Vernon, 1966). Transfer costs vary depending upon the structural characteristics mentioned above. In other words, transfer costs tend to be lowest for wholly owned subsidiaries, higher for independent licensees, and still higher for government enterprises in command economies such as China or the Soviet Union. Training costs also vary by industry (Stewart and Nihei, 1987; Teece, 1976).

Issues such as control and profit maximization as well as ownership rights dominate technology transfer case studies. Other relevant topics include the minimizing of risk and uncertainty, limiting training and consulting, and controlling managerial positions. It has been noted that countries in the first stages of development must depend largely on direct investment for technology transfer to occur. These countries are most affected (mostly to their disadvantage) by control and ownership restrictions. These countries also lack the absorptive capacity to advance mainly through independent licensing (Stewart and Nihei, 1987). As a country advances, case studies reveal that effective laws in support of patents and industrial property rights can have a major effect on the type and amount of technology that may be transferred.

Summary

In summarizing the current research on technology transfer, it is clear that there are numerous emphases in the extant research literature. In general, three trails of research have been identified; however, the research boundaries are not rigid. Research on technology transfer processes tends to be confounded with case analytic studies, and research on absorptive capacities tends to be highly diffuse. It is clear that considerable definitional imprecision clouds the research literature, which exacerbates measurement issues. Very little rigorous research exists, either of an empirical or a case-based nature, despite the abundance of one-shot case studies. The problems inherent in developing taxonomic categorizations of technology transfer then can only be resolved by significant attention to rigorous detail in studies involving technology transfer. Towards this end, we present in the next section, a research framework for analyzing technology transfer and apply it to the case of U.S.-China technology transfer.

A RESEARCH FRAMEWORK FOR ANALYZING INTERNATIONAL TECHNOLOGY TRANSFER

Elements of Technology Transfer

In an attempt to make sense of the various research trails, we specifically focus on the basic or core elements of technology transfer. Accordingly, Frame (1983) and McIntyre (1986) define the international technology transfer process in terms of six constitutive elements:

1. *The transfer item and its characteristics.* What is being transferred? Is it a tangible product, a process, or know-how or a combination of these things? Is the technology advanced or relatively old, government-controlled or uncontrolled? Each of these dimensions is important in classifying the relative sophistication of the technology transfer process and the demands it places on the technology supplier as well as the recipient.

2. *The technology supplier or donor.* Is the entity that possesses ownership rights to the technology public or private? What are the supplier's objectives, and how willing is it to make the technology accessible to others? What are the policies of the donor's government regarding international flow of technology to the recipient's country? These are but some of the dimensions useful in characterizing the motives of the technology supplier.

3. *The technology recipient or purchaser.* Is the acquiring entity public or private? What are its objectives in acquiring the technology in question? What resources does the recipient possess to finance and assimilate the technology transfer? What are the policies of the recipient government with regard to offering legal protection to proprietary technology?

4. *The transfer mechanism or channel.* How will the technology be transferred from the supplier to the recipient? Will the mechanism be one of licensing, equity joint venturing, or direct purchase? What will be the legal and managerial modalities governing the transfer?

5. *The rate of diffusion or acquisition.* How long does it take for the technology to reach the recipient and then be incorporated into the recipient plant, economy, and society? If it is a dynamic technology transfer agreement, how effectively can the recipient keep pace with technology refinements? At what rate does the technology reach the recipient's infrastructure, for example, through local sourcing of components that may be imported during the early stages of the transfer process?

6. *The absorptive capacity of the recipient.* What are the constraints, in terms of scientific, technological, and manufacturing infrastructure, that affect the recipient's ability to absorb advanced technology? This notion refers to technological capacity rather than the recipient's financial resource capacity covered in item 3.

The six elements of technology transfer listed above are interrelated. Each raises a number of strategic, political, and managerial issues in the context of international technology transfer. In the case of U.S.-China technology transfer, these elements are also uniquely idiosyncratic. The two countries differ greatly in terms of technical specifications, standards, culture, language, and infrastructure. Moreover, they also differ sharply in terms of short-term goals. Chinese recipients, for example, have a distinct preference for disembodied, advanced technology, which is also the most sensitive type of technology transfer to a communist bloc country as far as the U.S. government is concerned. There also appears to be a major discrepancy between the desired rate of technology acquisition on the part of the Chinese recipient and its ability to assimilate technology, as perceived by the U.S. supplier.

As our case studies reveal in much greater depth, there are numerous other equally significant differences that characterize U.S.-China technology transfer. These differences contribute to its uniqueness and also serve to illustrate the relative inapplicability of existing technology transfer models to the U.S.-China context. Instead, what is needed is a systematic, in-depth, and comprehensive exploration of specific U.S.-China technology transfer projects that can serve as a base for inductive theory building as well as practical managerial advice.

Research Questions

We began our research with two basic questions:

1. What are the critical issues confronted by U.S. companies transferring technology to China? And by Chinese companies acquiring technology from the United States?

2. What are effective and ineffective strategies for managing these issues?

The primary impetus for our questions came from problems that many U.S. companies have consistently reported in practice (Campbell, 1986, 1988). Based on preliminary interviews with U.S. executives and the experience of the senior author as science advisor to the U.S. Embassy in Beijing between 1980 and 1982, we observed that even companies with established track records of transferring technology to other developing countries confronted seemingly insurmountable problems in China. Hurdles, triggered by infrastructural, cultural, and communication barriers, arose at each stage of the technology transfer process, and most of the U.S. executives found that their earlier experiences had not prepared them for these problems.

To make our research as relevant as possible to an audience of senior managers, we made several policy decisions in the early stages of our research project:

1. *Focus on proprietary knowledge obtained through commercial transactions.* Technology transfer can refer to the acquisition of free knowledge, such as that obtained in scientific journals and government publications, or to proprietary knowledge, which is usually protected by patents and trade secrets (Frame, 1983). Proprietary knowledge, particularly that possessed by business entities, is usually transferred for financial consideration. Such transactions raise issues that may be quite different from the transfer of technology by an international government agency, for example. Our study focuses exclusively on the transfer of proprietary knowledge by U.S. business enterprises to Chinese plants.

   In particular, we concentrated on the transfer of manufacturing and design technology related to tangible products. Thus, we were not interested in studying the transfer of service-based technology, such as The Great Wall Sheraton Hotel, even though it was one of the earliest U.S.-China technology transfer agreements. It was our belief that we would observe the most complex interplay of technical, commercial, and cultural forces in the transfer of manufacturing and design technology. Moreover, by far the majority of U.S.-China technology transfer agreements were in the domain of manufacturing and design technology, thus also greatly enhancing the relevance and general applicability of our research findings.

2. *Concentrate on the study of licensing agreements and joint ventures.* A preliminary examination of existing U.S.-China technology transfer agreements revealed that licensing agreements are a very frequently used mechanism for U.S.-China technology transfer. This is in contrast to the common lore that "joint ventures are it." Joint ventures have historically been promoted by the Chinese policy makers because they want to increase the level of hard currency investment in the country. Many U.S. companies, on the other hand, view licensing as a less risky entry strategy that provides a foothold while preserving the option of equity investment at a later time. This appears to be particularly true in industry segments

that are highly capital intensive, such as heavy equipment manufacturing. As a result, we decided that we would study technology transfer agreements representing both licensing mechanisms and joint ventures.

3. *Obtain multiple stakeholder perspectives.* A full understanding of U.S.-China technology transfer could only be obtained through data representing multiple perspectives both in the United States and China. From the U.S. perspective, we interviewed representatives of the National Council of U.S.-China Trade (NCUST), renamed U.S.-China Business Council (USCBC), in Washington, D.C., as well as representatives at the U.S. Embassy in Beijing. USCBC is a private organization devoted to the development of business in China whose on-site briefings are tailored to organizational needs, market surveys on industrial sectors, trip arrangements and escort services, direct mail marketing, technical seminars, receptions and banquets, organization and statistical profiles, sales profiles, and negotiating teams and assistance in dispute resolution.

The stakeholders to a U.S.-China technology transfer project on the Chinese side are numerous and include representatives from the relevant ministries in Beijing, representatives from the provincial authorities, and, of course, the engineers and managers of the recipient enterprise. Each of these stakeholders is involved in a fairly central way, particularly during the early stages of a project. We therefore planned a series of interviews with each of these stakeholder groups.

The broad scope of our data collection is shown in Figure 1.1, a schematic of the key stakeholders in a typical U.S.-China technology transfer project.

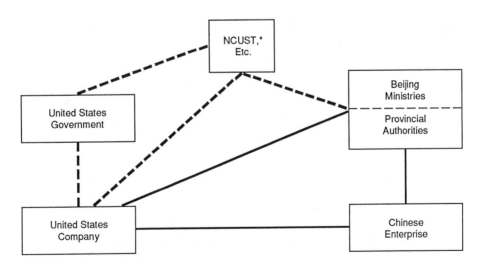

Figure 1.1 Key stakeholders in United States-China technology transfer.

4. *Select specific research sites.* We started our data collection in 1985 with an archival search of existing U.S.-China technology transfer agreements. In line with our research goals, we eliminated projects that focused exclusively on the service sector, such as hotels. We also, for the same reason, chose to eliminate oil drilling ventures. As previously noted, the remaining list consisted mostly of licensing agreements and joint ventures involving the transfer of core manufacturing and design technology.

From this list, we selected those companies that could be termed the pioneers in U.S.-China technology transfer. These companies were the early entrants and had typically signed their agreements in the 1981–1982 period. Consequently, by 1985 these companies had already had several years of experience in the actual "nuts and bolts" of transferring technology to their Chinese partners. Since our goal was to adopt an integrative approach, we intended to collect data on experiences during all stages of the technology transfer project, namely, the early contacts, negotiations, start-up, and the management of the ongoing process.

Starting with a list of fewer than 100 companies that met our criteria, we invited the chief executive officer of each company to participate in our research program. In exchange for providing the research team entry, we offered to share our preliminary findings with the company after our interview phase was complete.

After considerable elimination of eligible firms, four companies agreed to participate: (1) Foxboro Company: control instrumentation technology, joint venture; (2) Westinghouse Electric Corporation: turbine and generator technology, licensing; (3) Combustion Engineering: boiler technology, licensing; and (4) Cummins Engine Company: diesel engine technology, licensing.

5. *Use a clinical case strategy for collecting data.* We chose to rely on in-depth case studies of a relatively small number of individual technology transfer agreements as our strategy of collecting data. In view of the relative paucity of meaningful, integrative U.S.-China technology transfer research, that is, research combining macro-policy issues with micro-management issues, we concluded that an exploratory data collection strategy to capture firsthand the rich experiences of senior U.S. and Chinese managers and engineers would be most appropriate.

Most previous research on U.S.-China technology transfer attempted to study a large sample of companies (e.g., Campbell, 1986). As a result, cross-sectional data on the types of technology transferred, the investment patterns, and so forth were reported. However, we were more interested in the "how" of technology transfer rather than a first-order understanding of the "what." By "how" we mean the patterns in the interaction among the U.S. executives and their Chinese counterparts, among technical and managerial personnel, as well as among the U.S. executives within the U.S. companies. By using cases to explore these interactions, we hoped to capture a richer understanding of the dynamics of U.S.-China technology transfer.

Stages in Technology Transfer

In taking a retrospective look at international technology transfer projects, we observed that these projects passed through four reasonably discrete stages: (1) Development; (2) negotiations; (3) start-up (implementation); and (4) management of the ongoing process. (See Appendix for a more detailed outline.)

During stage 1, Development, executives essentially went through an exploratory phase. Executives from the supplier company assessed the potential of the recipient's market, obtained preliminary information on various potential partners, explored U.S. government licensing and other constraints, and so forth. Similarly, during stage 1 representatives from the recipient organization tried to assess the reputation of the supplier's technology, its appropriateness for their context, and so on.

During stage 2, the negotiations stage, the two companies concentrated their efforts on reaching a formal agreement. Details of the precise product and technology to be transferred, royalty payments and other commercial considerations, and alternative mechanisms of technology transfer were resolved during this phase. If an agreement was successfully reached, the stage was set for start-up of the actual technology transfer process.

During start-up (stage 3), the two enterprises implemented the actual transfer of technology. The specifics depended on the nature of the technology and the type of agreement but it usually involved the transfer of equipment, technical documentation, and training. Once the start-up phase was over, the supplier and recipient settled into stage 4, management of the ongoing process. This stage included an ongoing process of monitoring and feedback. New technologies could be added as the relationship evolved.

Data Collection

We were interested in understanding the dynamics during each of these stages. We therefore developed a comprehensive interview protocol based on the five stages (see Appendix) that listed a series of questions focusing on key issues at each stage of U.S.-China technology transfer. Moreover, we refined the questions and issues as the interviewing progressed within a company. Armed with this protocol and background archival data on each company, the first phase of our data collection concentrated on interviewing executives from the four U.S. companies in the United States.

Each company assembled a group of engineers and managers who had been involved with the particular company's China project since its inception. We interviewed between six to ten individuals per company in the first round.

A typical interview lasted two hours and was tape recorded. The tapes were later transcribed and used as a basis for preparing an in-depth case study. The four-stage outline in the interview protocol was also used as the structure of each case. Hence, each case describes the company's experiences in the developmental stages, during negotiations, during the start-up, and during the ongoing process. A section on future plans is also included.

Since the process of preparing the case study raised several questions and indicated a need for additional data, we revisited each company or visited another facility of the company for a second round of interviews. Following the second round, each case was revised and submitted to the company for verification. In all cases, the companies suggested only minor corrections.

The next major round of data collection occurred in China. Interviews were planned in China through two channels of communication. First, executives from the U.S. firms contacted their expatriate managers in China to request their participation in our research. This occurred in the cases of Foxboro, Combustion Engineering, and Westinghouse. At the time of our field research in China, Cummins Engine Company did not have an expatriate in China.

A more important channel for establishing research access in China, however, was the contact made by the senior author directly with representatives of the National Research Center for Science and Technology for Development in Beijing, the think tank of the State Science and Technology Commission. Through these contacts, meetings were scheduled with high-level representatives from the Ministry of Machine Building Industry (MMBI) in Beijing, more recently reorganized as the Ministry of Machine Building and Electronics Industry, the recipient plants in Harbin and Shanghai, and with the provincial authorities in Shanghai. In Shanghai, the research team benefited from the organizational services of the Economic, Legal and Social Consultancy Center of the Shanghai Academy of Social Science.

The interview protocol we followed in China was identical to the one used in the United States, with only minor differences. A key difference in the interview process in China, however, was that we met with large groups of officials from relevant ministries instead of with individual executives. Thus, for example, we had three days of intensive interviews at the Ministry of Machine Building Industry, during which about 30 MMBI officials were present in the room. These officials included individuals who had been directly involved in the early contacts, negotiations, and start-up of each of the technology transfer projects in our study. At the recipient plants, we interviewed the chief engineer or deputy chief engineer and the vice manager in one case (Chongqing Automotive Engine Plant), who in all cases had been actively involved in the technology transfer from the beginning.

We did not tape record our interviews in China. Due to the need for language interpretation and also due to the sensitivity of the Chinese interviewees, we decided to rely on handwritten notes. Since all three members of the research team were present during all interviews in China, except in Chongqing, it was relatively easy to coordinate note-taking, interviewing, and questioning within the team. Following an approach similar to what we followed with developing case studies on the U.S. companies, we developed somewhat shorter case studies describing the Chinese perspectives of each of our technology transfer projects. Here again we adopted our stage model as the central outline for organizing each case study. Each case was sent to the relevant Chinese participants for review. Again, we received only minor corrections.

In Part II, we present the case studies that we have prepared using the approach outlined above for four projects. The case study of the U.S. company is followed by a case study of the Chinese affiliate(s) involved in the same project to facilitate a comparison of the two different perspectives.

CHAPTER 2 _____

Environment for Technology Acquisitions in China

BACKGROUND

In the early years following the foundation of the People's Republic of China in 1949, foreign trade was primarily limited to the USSR and East European satellite countries. On a restricted scale, technology imports from Western Europe and Japan took place in the early 1960s, but beginning about 1966 the isolationism fostered by the Cultural Revolution strictly limited contacts with foreign countries. As the Cultural Revolution receded after 1969, foreign trade was resumed on a limited scale. It began to accelerate after 1971, and by 1978 foreign trade reached an annual volume in excess of U.S. $20 billion. China's foreign trade has grown by leaps and bounds since then, reaching $102.9 billion in 1988. An overview of Chinese foreign trade statistics is presented in Table 2.1.

This chapter presents an overview of the environment in which technology acquisition takes place in China. We shall discuss a series of key issues affecting the primary participants in technology transfer from the United States to China.

The historic events leading to the explosive growth of China's commercial contacts with the outside world and, in particular, with the capitalistic countries have been described by Ho and Huenemann (1984). We shall briefly review them here.

The unchallenged leader of the Chinese Communist Party since the end of the Long March in 1935, Mao Zedong (Mao Tse-tung), died in 1976, and

TABLE 2.1 China's Foreign Trade Statistics

		$ Billions		
Year	Total Trade	Export	Import	Trade Surplus
1950	1.13	0.55	0.58	− 0.03
1952	1.94	0.82	1.12	− 0.30
1957	3.10	1.60	1.50	+ 0.10
1965	4.25	2.23	2.02	+ 0.21
1970	4.59	2.26	2.33	− 0.07
1973	10.98	5.82	5.16	+ 0.66
1975	14.75	7.26	7.49	− 0.23
1978	20.64	9.75	10.89	− 1.14
1980	37.32	18.27	19.55	− 1.28
1982	41.60	22.32	19.28	+ 3.04
1984	53.55	26.14	27.41	− 1.27
1985	69.60	27.35	42.25	− 14.90
1986	73.85	30.94	42.91	− 11.97
1987	82.65	39.44	43.21	− 3.77
1988	102.8	47.5	55.3	− 7.8

Source: The data for 1950–1987 are from *Zhong'guo Tongji Nianjian, 1988 (China Statistical Yearbook, 1988),* ed. State Statistical Bureau (Beijing, China: China Statistical Publishing House, 1988).

The data for 1988 are from the Communique on Statistics of National Economic and Social Development, State Statistical Bureau, for 1988.

The data represent PRC customs statistics for the years 1982–1988. Data for 1950–1980 are published by the Ministry of Foreign Trade.

the isolationist "gang of four" and their followers were purged before the year was out. The elders of the revolution turned for leadership to Deng Xiaoping, a seasoned administrator who had been purged twice from prominence during the preceding decade. During the years 1977 and 1978 the policy of opening up to the outside world (*kaifang zhengce*) was planned, and the new strategy for the country's development was officially announced at the Third Plenary Session of the Eleventh Central Committee of the Party. It was decided to expand economic cooperation with other countries on terms of "equality and mutual benefit."

After decades of quasi-isolation, the Chinese leadership recognized the necessity to learn from the advanced industrialized countries of the West and Japan. By 1978, it became known from pronouncements by members of the elite, that China would now be prepared not only to buy technology but also to accept investment from the developed countries of the capitalist world. These pronouncements stirred the imagination of organizations the world over. In July

1979 China promulgated the new "Law of the People's Republic of China on Joint Ventures," which spelled out the basic regulations for foreign investment in Chinese industrial enterprises. This was a near-revolutionary step for the country in view of its recent history of isolationism and the glorification of the doctrine of self-reliance (*zili gengsheng*).

ECONOMIC ENVIRONMENT AND POLICY

Political Priorities

China's economic policy since the establishment of the People's Republic in 1949 had centered around three central issues. First, the orthodox party line, supported by Mao Zedong, maintained that the fundamental problem of China centered around the transformation of the productive relations, or the development of socialism. It was believed that once this was accomplished, economic construction would progress rapidly. Economic construction per se was, therefore, not considered to be the first priority. True, there were leaders, such as Deng Xiaoping, who did not agree with this approach, but they were effectively kept in check until 1976.

Second, the radical party stalwarts also supported a Stalinist economic policy that favored an unbalanced growth strategy, that is, high investment rates with priority allocations going to the capital goods sector. It was believed that the rate of economic growth of the country was determined by the size of the capital goods sector, and the consumer goods sector was to be expanded only as long as it did not compete with heavy industry. Also, it was held that production could be expanded through political mobilization or by motivating the workers by means of political propaganda. Opponents of this view believed that unbalanced growth would result in a waste of scarce resources and would produce strong inflationary pressures. This faction, in the position of powerless opposition until Mao's death except for a short period between 1972 and 1974, advocated balancing the growth of the capital goods sector with the consumer goods sector, which includes light industry and agriculture.

Third, the faction in power until 1976 practiced directive planning by the central government. Their opponents, however, advocated the use of economic levers at the macro or state level (e.g., pricing, taxation, and subsidies) to regulate the economy, some decentralization of authority to the provincial and local levels of government, and the granting of a degree of autonomy to the enterprises. The degree to which these reforms are appropriate to a socialist economy, and in particular to China, has been the subject of intense debate which continues to this day.

After the downfall of the Gang of Four, the Chinese leadership shifted priorities from the development of socialism to economic construction and the "Four Modernizations"—agriculture, industry, defense, and science and technology.[1] The Fifth National People's Congress (NPC) at its First Session in early 1978 adopted the Ten-Year Plan for 1976–1985. The plan was over-ambitious, retained the principle of unbalanced growth, and repudiated isola-tionism. Eleven months later, in December 1978, the Third Plenum of the Eleventh Central Committee discarded the Ten-Year Plan and sanctioned the strategy of balanced growth. However, it was decided that an interim period of "readjustment, restructuring, consolidation, and improvement" was required to correct economic imbalances created by past policies. During this period, development was to be slowed. In fact, some recently signed contracts for pur-chases from abroad were postponed or canceled, to the chagrin of a number of Japanese and Western companies (Ho and Huenemann, 1984).

Economic Goals and Reforms

In December 1981 and in his report to the Twelfth Party Congress in September 1982, then Premier Zhao Ziyang announced the aim to quadruple the Gross Annual Value of industrial and agricultural production by the year 2000. Analysts characterized the plan as ambitious but feasible. In addition to the past imbalance between heavy and light industry, it was concluded that the investment within the heavy industry sector had been unbalanced to the detri-ment of the development of energy resources. In October 1982 Premier Zhao announced that energy would be a major bottleneck in the country's economic development plans. He estimated that China probably could only double its energy output by the end of the century and therefore the industrial plant had to be modernized and made more energy efficient if the Gross Annual Value of production was to be quadrupled. World Bank studies also agreed that energy would be the most serious constraint on China's economic growth during the 1980s.

In his report to the Fourth Session of the Fifth NPC in late 1981, Premier Zhao Ziyang advocated qualitative change of the means of production through technological innovation and improved management to supplement mere quantitative change or the building of new plants. Existing plants were to be renovated and equipped with modern machinery and imported technology in order to increase their productivity. Also, the past stress on heavy industry at the expense of consumer goods and the lack of emphasis on raising the stand-ard of living of the people was said to have been harmful to morale and motivation.

In November 1982 China's Sixth Five-Year Plan for the years 1981–1985 was publicized and followed the guidelines laid down by the Party Congress held in September. It reaffirmed the government's commitment to readjustment, to reform, and to the balanced development of the economy. Priority was to

---

[1]The Gang of Four refers to Jiang Qing (Chiang Ching), Chairman Mao Zedong's widow, and three of her associates, who attempted to take control of the government after Mao's death on September 9, 1976. They were arrested in October 1976.

be given to the development of the energy and transportation sectors, which together were to receive 38.5 percent of the total investment in capital construction during the plan's five years. Total fixed investment was budgeted at 360 billion yuan (about U.S. $200 billion). Education, science, culture, and public health were to receive 68 percent more than during the preceding five years (1976–1981). The gross value of industrial and agricultural output was projected to rise at an annual rate of 4 percent, and per capita consumption projected to increase 4.1 percent per year, was to keep in step with this rate.

Reforms were aimed at modifying past practices in three areas. Incentives and rewards for productivity would now take the place of the emphasis on absolute egalitarianism. Controls over directive planning and administrative matters, previously concentrated at the level of the supervising ministries, were to be loosened. Enterprise autonomy and decision-making power at the enterprise level were to be increased. The reforms were initially introduced on an experimental basis in six enterprises in Sichuan Province in October 1978. Within two years they were expanded to include 6,600 key enterprises producing a total of 60 percent of the country's total industrial output.

According to the reforms enacted, once an enterprise had fulfilled its quota under the plan, it was entitled to market in China or abroad any excess product that was not bought by the state marketing agency. Also, each enterprise was empowered to retain a percentage of its after-tax profits to be used for workers' welfare programs, for bonuses paid to the workers, and for enterprise development. Unsatisfactory workers could be disciplined or, in extreme cases, dismissed, according to official pronouncements. It was also announced that unprofitable enterprises would be pressured to close down, merge with others, or change their direction (*guan, ting, bing, zhuan*). In the face of resistance from less successful enterprises and less developed regions as well as from members of the leadership who feared the loss of state revenues, the reforms were modified to proceed at a slower pace during a time period to extend into the Seventh Five-Year Plan (1986–1990). Price reform was also postponed to 1986 or beyond (Ho and Huenemann, 1984).

Following a period of uncertainty as to which side, the reformers or their opponents, would gain the upper hand, it appeared after the Thirteenth Party Congress in October 1987 and the Seventh NPC in March 1988 that the reform faction had succeeded in gaining full control of the decisive party and government bodies (Frankenstein, 1988). Nevertheless, the rate at which some of the more radical reforms would be instituted remained uncertain. Price reforms were enacted to some degree in 1987–1988, but resulted in an inflation rate, estimated to exceed 25 percent per annum. Retail prices were reported to have increased by 18.5 percent during 1988, and the consumer price index rose by 20.7 percent (*Beijing Review,* 1989). These statistics were characterized as "figures never witnessed since the founding of New China in 1949." The rate of inflation was also discussed at a press conference in January 1989 by a spokesperson of the State Council, who put the inflation rate at 17.7 percent for 1988 (Beijing Television Service, 1989). He also acknowledged that many Chinese have experienced a higher rate because in their daily life they purchased only a part of the range of commodities on which the officially published inflation rate is based.

In order to curb inflation and the excessive growth in the overheated economy, a policy of retrenchment was instituted in September 1988. This policy was to last for two years (New China News Agency, 1988a), and during this period the government would seek to check demand by restricting bank loans and money supply (Zhang and Luo, 1989). A *New York Times* article reported data released by the State Statistical Bureau (Ignatius, 1989): Inflation reached 26 percent in December 1988 and remained the same for January 1989; however, in most urban areas it was 31 percent, according to official figures. Furthermore, inflation outpaced income gains for 35 percent of urban families during 1988. According to this report, the Chinese authorities pushed rapid price reforms to allow most prices to rise to market levels in early 1988. However, the ensuing inflation caused so much dissatisfaction that it was announced that price reforms would be postponed for two years. Also, a policy of reforming the system of ownership of state enterprises by selling shares has been postponed for two years except in isolated experimental areas.

The State Statistical Bureau also warned that the state was encountering difficulties in its attempts to control inflation. The measures instituted in September 1988 produced mixed results, although the industrial growth rate was successfully reduced from 18 percent in December 1988 to 8.2 percent in January 1989 (Ignatius, 1989). Possible negative effects of the retrenchment policy on the economy are discussed by Yang (1989), who predicted that it may result in recentralization and increased bureaucratic controls with the potential for arbitrary administrative implementation and excessive emphasis on *guanxi* (personal connections).

The significant lowering of the purchasing power of a majority of the urban population, caused by the inflation that followed price reforms, is believed to have been an important factor contributing to the outbreak of unrest during April-June 1989. Students initiated the demonstrations but many professionals and industrial workers joined in the protests against restrictive controls on the media and the alleged corruption among officials and their relatives. The subsequent forceful suppression of the protest movement by military force on June 3–4, followed by widespread arrests and executions, marked a return to political conservatism, reminiscent of the Cultural Revolution era. However, official pronouncements consistently pledged the government's continued adherence to a policy promoting Chinese economic cooperation with foreign countries.

Technical Work Force Resources

Before considering the extent of the science and engineering work force of China, we examine some indicators of the college-trained population in general. Official publications of China's State Education Commission tabulate the number of university graduates by year and by major subject. The graduates are also subdivided according to two types of courses: the full-term courses of four years' duration and the short-cycle courses that train higher level technicians and are two to three years long (Orleans, 1987; Schnepp, 1988).

Cumulatively over the period 1949–1987, the number of full-term university graduates totaled 2,890,000. However, in the year 1987 alone, 253,800 students graduated, indicating that the capacity of Chinese universities has been greatly expanded during the past eight- to ten-year period and that every class of graduates significantly affects the existing pool of university-trained work force. The World Bank has carried out extensive studies of China's education system and has reported that the "enrollment ratio" in institutions of higher education, relative to the age cohort, was 4.8 percent compared with 8.1 percent for India, 10.3 percent, for Mexico, and 12.5 percent for Brazil (World Bank, 1986). In 1985, the median for this indicator was 3.8 percent for the less-developed countries (LDCs). China was reported to aspire to reach the LDC upper quartile of this indicator by the year 2000. In 1985 the LDC upper quartile stood at 10.5 percent, and therefore it will take exceptional efforts for China to reach its ambitious goal. On the other hand, in terms of absolute numbers, China's yearly crop of university graduates is very reasonable, ranking just below Japan and Brazil and above South Korea, the United Kingdom, and both Germanies, according to United Nations Educational, Scientific, and Cultural Organization (1986) and the National Science Foundation (1985), which tabulate the number of first degrees conferred in selected countries.

We now examine the technical work force resources. At the end of 1987, China had a total of 1,150,000 solidly trained engineers who were graduates of four-year, university-level programs. These engineer graduates accounted for well over half of the full-term graduates in all fields of science and engineering or about 40 percent of the full-term university graduates in all subjects. In addition, about 200,000 more engineers graduated from shorter (two- to three-year) college-level courses, qualifying them to be high-level technicians. About 1,850,000 technicians graduated from technical high schools. These numbers do not include about 500,000 engineers who graduated from universities during the tumultuous Cultural Revolution from 1966 to 1976 and whose qualifications are not as well established. At the same time, the number of graduates from four-year courses in the natural sciences was close to 270,000, nearly 10 percent of the total, not counting the approximate 100,000 Cultural Revolution graduates. The number of short-cycle and high school technician course graduates in these subjects was negligible.

In order to assess China's work force resources in science and engineering relative to other countries, comparisons can be made with tabulations made by UNESCO (1986). The number of scientists and engineers with solid educational background is impressively close to that in countries considered part of the industrialized world: China ranks above all the developing countries and above East Germany (DRG) and closely below West Germany (FRG). When, however, the absolute numbers are converted to figures relative to population (number of scientists and engineers per million of population), China places considerably lower than these nations. In these terms, China ranks considerably below Brazil, Argentina, and, of course, all developed countries.

By the end of 1987, China had about 60,000 scientists and engineers who had completed graduate studies, most of whom were considered to have reached a level equivalent to an M.S. degree. Only 3,500 to 4,500 of these scientists hold Ph.D. degrees, that is, have had training in innovative research. In addition, in 1985 there were "87,000 high level specialists in the natural sciences [of whom] more than half were over 55 years of age and 30 percent were 61 or older," according to the State Science and Technology Commission (1987, p. 316). As a result, the level of achievement to be expected from the R&D-employed work force in the near future is limited. In absolute numbers, however, the number of workers employed in the R&D sector compares favorably to those in other countries.

In summary, all parties, Chinese government and outside analysts, agree that China has a shortage of skilled workers and needs to expand its "entire system of higher education" by a factor of about five, according to World Bank (1986). Conventional university enrollment was projected to grow at an annual rate of 10 percent for 1983–1990, according to State Education Commission sources. Alternative educational delivery systems have been incorporated in the plans, including the expansion of short-cycle technical college courses and various forms of adult education. There has been a great demand for higher and specialized education, and the increase of trained work force was, until recently, strictly limited by availability.

In recent years Chinese youth have become somewhat disenchanted with academic or other specialized education as inflation has impinged negatively on the "intellectuals" who are mostly employed in educational and research institutions and whose real income has fallen behind that of factory workers and taxi drivers. The government is making great efforts to expand university and continuing education enrollments and has experimented with nonresident students and other means to increase the population of existing facilities. The results have been mixed, and students have, on occasion, protested against overcrowded conditions.

The number of available personnel is certainly one important factor determining a country's productive capacity and ability to compete in international markets. However, many other factors determine how effectively the available work force is used. A great deal remains to be done in China to raise the level of motivation and to establish a suitable system of rewards for scientists, engineers, and technicians. First and foremost, graduates must be allowed to participate in the job selection process and personnel must be able to improve their employment situation when desired. Recently, reforms have been introduced to gradually make the students themselves responsible for finding jobs. Several universities have been designated to implement the new policy on an experimental basis. A complete transformation of the graduate placement system is hoped to be accomplished by 1992.

Experts agree that China's greatest need is to support or replace, whenever possible, fully qualified engineers and scientists with technicians. To this end, the ratio of short-cycle technician graduates to full-term university graduates has been on the increase in recent years. China also needs to improve the competence of its research and development work force if it wishes to eventually assume leadership as an industrially developed country in some areas of technology, rather than forever follow others. One important source of the necessary highly trained research personnel is the pool of Chinese students who are studying for Ph.D. degrees in industrialized countries, including the United States. Several hundred of these students have already returned to China from Western Europe, but only a few have returned to China from the United States. The return of these students to China has been a matter of great concern to the Chinese authorities in recent years. Following the events of summer 1989, it is anticipated that most Chinese students will try to remain in Western countries indefinitely.

FOREIGN TRADE RELATIONS

As already mentioned, the new feature in Chinese plans for development and modernization was the readiness to seek and to accept loans from foreign and capitalistic sources, including banks, companies, and international agencies, such as the World Bank. The new rules under which foreign capital could be invested originally included five basic types of "special arrangements" (Ho and Huenemann, 1984), with a sixth added later:

1. Processing and assembling (*lailiao jiagong, laijian zhuangbei*).
2. Compensation trade (*buchang maoyi*), including direct and indirect compensation trade.
3. Joint ventures (*hezi jingying*) are equity joint ventures administered in accordance with the Joint Venture Law.
4. Cooperative ventures (*hezuo jingying*) are cooperative projects not conforming with the Joint Venture Law and are therefore more flexible. These ventures are governed by the contract between the parties. Examples are co-production or other joint manufacturing projects.
5. Cooperative development (*hezuo kaifa*), under which foreign companies in cooperation with Chinese enterprises engage in exploration for and development of natural resources. Examples are off-shore oil and gas exploration and coal mine development.
6. Wholly-owned foreign enterprises (*duzi jingying*) were added later, with relevant regulations published in January 1986. The first well-publicized, wholly-owned foreign enterprise outside the Special Economic Zones (discussed later in chapter) was established by Minnesota Mining and Manufacturing (3M) Corporation in November 1984 (Marshall, 1987).

TABLE 2.2 U.S.-China Foreign Trade Statistics

	Billions U.S. $			
Year	Total Trade	Export to China	Import from China	Trade Gap
1983	$4.65	$2.17	$2.48	$0.31
1984	6.38	3.00	3.38	0.38
1985	8.08	3.86	4.22	0.36
1986	8.35 ($7.35)	3.11 ($4.72)	5.24 ($2.63)	2.13
1987	7.87 (7.87)	3.04 (4.83)	6.83 (3.04)	1.79
1988	9.97	3.34	6.63	3.29

Source: Data for 1983 are from *China Business Review,* 15, 3 (May-June 1988), p. 57. Data for 1984 to 1988 are from *China Business Review,* 16, 3 (May-June 1989), p. 45. Data in parentheses are from *Zhong'guo Tongji Nianjian, 1988 (China Statistical Yearbook, 1988),* ed. State Statistical Bureau (Beijing, China: China Statistical Publishing House, 1988), based on Chinese customs statistics.

Note: Data from U.S. sources differ from Chinese customs statistics, in that the latter exclude trade between China and the United States via Hong Kong.

The U.S.-China trade volume has kept pace with the general increase in China's overall trade (Table 2.1) and is shown for recent years in Table 2.2. China's imports from the United States have undergone modest growth, but its exports have increased substantially to widen the trade gap to $4.0 billion in 1988. Chinese customs statistics, given in parentheses in the table, do not include trade via Hong Kong and, as a result, do not reflect the trade gap adverse to the United States.

Ho and Huenemann (1984) summarized the open-door policy in its original form as having five key features. Three of these features—the concern for effective technology transfer from abroad, the thrust to expand exports, and the solicitation of foreign investment to finance development and modernization—were considered to be essential elements of the policy. Two other features—special emphasis on energy resource development and priority for the development of the coastal regions (discussed later in chapter)—were considered not to be basic elements of the open-door policy but were expected to persist since they were important to the Chinese economy in their own right. Simon (1987) added to these elements of the open-door policy the aim to effect changes within the Chinese industry, such as the development of horizontal subcontracting relationships and the improvement of overall management.

Chinese Infrastructure for Foreign Trade

In the early years of "opening up," contacts with foreign firms were limited to only 16 foreign trade corporations (FTCs) (Frankenstein, 1988), which were subordinate to the Ministry of Foreign Trade, the forerunner of the Ministry of Foreign Economic Relations and Trade (MOFERT). Two of the best known of these FTCs were the China National Technical Import Corporation (Techimport), which was responsible for technology and complete plants import, and the China National Machinery Import and Export Corporation (Machimpex), which dealt with the import of a large variety of machinery and equipment. Beginning in 1979, industrial ministries and provincial and local governments were empowered to establish their own FTCs, and the branch offices of the central FTCs were invested with increased authority. By 1988 there were as many as 1,200 organizations authorized to negotiate with foreign companies. In November 1987, as part of the trend toward deregulation, the FTCs were made responsible for their own finances and operations and were allowed to retain a large part of their earnings (Frankenstein, 1988). These developments, although applauded in some quarters as a sign of decentralization of authority and decision making, caused some confusion for foreign businesspersons, as provinces competed fiercely for foreign capital and underbid one another in the process.

In principle, MOFERT retained the authority over foreign trade and had to approve all special and foreign-invested trade projects, but MOFERT delegated this authority to the provinces and the industrial ministries as long as two conditions were satisfied. First, the project concerned could not require central resources or impinge on the balance of energy resources, nor could it produce commodities that took up China's export quotas. Second, the project had to be below a defined limit of registered capital. For Guangzhou (Canton) city and Fujian Province, there was no limit; for Liaoning Province and Beijing and Tianjin municipalities, the limit was $5 million. Shanghai's limit was $10 million. In spite of these efforts at decentralization, most Chinese enterprises were still kept isolated from the foreign companies with whom they, in principle, had business dealings. This situation was not conducive to efficient technology transfer by information diffusion and feedback mechanisms (Ho and Huenemann, 1984).

The accelerating growth in Chinese foreign trade (see Table 2.1) occurred while a high level of central controls on commodities and capital remained in place. For example, in early 1982 the State Council imposed significant restrictions on imports in order to protect domestic industrial development. Also, during the years 1984–1987 there were repeated reversals of reforms previously implemented. For example, in 1984, 14 coastal cities were opened to foreign trade and investment but the list was cut back a short time later in view of bureaucratic opposition (Frankenstein, 1988). In 1988 the Chinese economy still largely remained a "command bureaucratic economy" in which the government continued to play a crucial role (Frankenstein, 1988), despite the trend toward greater openness and deregulation. In their speeches to the Thirteenth Party Congress and the NPC in 1987 and 1988, respectively, top leaders confirmed these trends.

The policy of retrenchment, which was enacted in September 1988 to combat an unacceptable rate of inflation, reversed the policy of decentralization even further. It became clear that the central government may utilize any measure it deems effective to return order to the overheated economy, including reinstituting central controls. However, Zhao Ziyang, the party general secretary during January 1987–June 1989, has stated that development will continue and the government's policy will not adversely affect foreign participation in China's economic development (New China News Agency, 1988d). In his words,

> It is an adjustment in the process of development. Though the scale of construction will be reduced and the speed of development will be slowed down, the scale will be comparatively large and the speed comparatively fast (p. 5).

In particular, Zhao emphasized that the Chinese government "will institute no restrictions for foreign entrepreneurs who wish to establish solely funded (wholly owned) enterprises in China" (p. 5). He also reiterated the preference for joint ventures that are self-sufficient in foreign exchange. Observers expect that the new review process for foreign-invested enterprises will primarily attempt to prevent the building of new hotels and other facilities requiring "capital construction," but that it will not affect the licensing of manufacturing joint ventures. Nevertheless, the availability of loans has been severely cut back; it is expected that the start-up of new enterprises will be hampered for the immediate future. It has been reported, for example, that 14,000 investment projects have been canceled or postponed and an unspecified number of foreign-invested enterprises are probably included in this number (Ignatius, 1989). Also, following the military suppression of the democracy movement in June 1989, cutbacks of foreign investment may be expected, since companies will judge the internal political and social situation in China to lack long-term stability.

Coastal Regions

The Chinese government policy of opening up to the world assigned a special function to the coastal provinces and municipalities. These regions were recognized as having the most experience in contacts with foreign countries; Liaoning in the Northeast (or Manchuria) has extensive background in dealing with Japan, and most other coastal regions have had contacts with the United States and West European countries. Beijing was considered on the same level of importance with Shanghai, Tianjin, and Liaoning because of its extensive resources, even though it has had less foreign trade experience. Also, these provinces are the most developed industrially; the four regions have the highest per capita and per worker Gross Value of Industrial Output (GVIO). Further, they together account for 28 percent of the country's industrial output, although they only have 6 percent of the population. In terms of foreign trade statistics, the four regions' total exports represented 45 percent of the country's total in 1980 (Ho and Huenemann, 1984). Also, many industrial enterprises designated for renovation under central government policy are located in these four regions.

Another issue relating to the coastal regions, although partly economic, also has important political implications (Ho and Huenemann, 1984). The coastal areas have a history of extensive contacts with "overseas Chinese," ethnic Chinese living abroad who came from these regions. The overseas Chinese communities were considered valuable potential sources of both expertise and capital. The most important such communities are located in Hong Kong, Macao, and Taiwan, and the strategy to build bridges between them and China dovetailed with political aims of unification of these areas with the mother land. Also, the policy of readjustment would be furthered preferentially by developing the coastal provinces and municipalities, since a large proportion of light industry is located in these regions. The coastal areas were granted a degree of autonomy in dealing with foreign companies and, as a result, local entrepreneurial initiatives have been widespread (United States/China Joint Sessions, 1988).

Based on the considerations noted above, Guangdong Province, adjoining Hong Kong and Macao, and especially Guangzhou city, and Fujian Province, opposite Taiwan, were given special status under the foreign trade regulations. Their preferential treatment under the decentralization of authority relative to foreign trade activity has already been mentioned. These two provinces were also granted greater autonomy in the management of foreign trade under a system of sharing foreign exchange revenues with the central government according to rules guaranteed to remain unchanged for the five years, 1980–1985. Among the special privileges were the power to pay higher wages and thereby to create incentives, the authority to approve special trade agreements, to control the province's exports and imports, and to use the province's own foreign exchange holdings (excepting commodities balanced by the central government) to borrow directly from foreign sources and to operate Special Economic Zones.

Special Economic Zones

Special Economic Zones (SEZs) were conceived in order to create designated areas where freer interaction with foreign investors would prevail and where market forces would be permitted to play a more prominent role. In the process, the efficacy of novel schemes for a more liberal environment for foreign-invested industry could be tested. The first SEZ, in Shekou, was authorized in January 1979, and the Xiamen (in Fujian Province) and Shenzhen (adjacent to Hong Kong in Guangdong Province) SEZs were established in October 1980. In mid-1982 laws were promulgated that codified the governance of the SEZs. Enterprises in the zones were to have greater flexibility in management and could offer greater incentives. A wide variety of business and industrial activity was encouraged, and foreign investment as well as wholly-owned foreign enterprises were welcomed. The primary aim was to promote the production of commodities for export; the sale of products in China proper was subject to the same regulations as imports from abroad, including the requirement for special licenses as well as customs duties and other taxes. The incentives offered included preferential taxation (a flat rate of 15 percent on earnings, compared with 33 to 50 percent in other regions), management independence, freedom to employ foreign nationals, and employment governed by contracts with flexible salaries and wage incentives. Since the SEZ administration was given authority over the local government agencies, red tape would significantly be reduced. The reasons for establishing the SEZs in Guangdong and Fujian, as mentioned above, included the presumed ability to tap into resources from overseas Chinese (both know-how and capital) and the desire to establish relations that would facilitate the reunification of the mother land with Taiwan (for Fujian) and Hong Kong and Macao (for Guangdong).

Unfortunately, the performance of the SEZs has been disappointing (Simon, 1987). Between 1979 and the end of 1985, the zones attracted $1.17 billion in foreign investment, about 20 percent of total foreign investment in China over the same period. Much of the funds went toward building up infrastructure, and only a small part of the authorized projects were production- and export-oriented. *Entrepôt* trade was successfully stimulated, but the "productive economy base" of the SEZs was not firmly established. At a meeting held in February 1986, the status of the SEZs was assessed; urgent changes in policy aimed at ensuring export orientation in the future were recommended. Such a change, however, required basic reorganization, since the products had been of uneven quality and had not been competitive on the world market.

Recent Patterns in Foreign Trade

Primary policy aims have remained unchanged since the early 1980s. They include the absorption of foreign technology and the promotion of import substitution, the upgrading of domestic products and promotion of their export into the world market, the further expansion of the country's foreign trade, and the conservation of foreign exchange (Ho and Huenemann, 1984; Frankenstein, 1988). The recognition of the country's urgent need to enlist the cooperation and financial support of foreign companies from industrialized and capitalist countries resulted in the promulgation of the Joint Venture Law in 1979. China also realized that it not only needed manufacturing and machinery technology but also management expertise and that such know-how was included in technology import plans. The establishment of foreign-invested joint ventures, it was believed, would offer the best opportunity for the acquisition and absorption of the entire range of technologies required. Implementing regulations for the Joint Venture Law, however, were only published in September 1983, indicating some ambivalence on the part of the authorities to proceed wholeheartedly with *kaifang zhengce*, opening up the economy to foreign participation (Simon, 1987).

Decentralization and reforms have had their negative results as well. Once controls were loosened, Chinese enterprises purchased a large volume of foreign products and technology; as a result, the country's foreign exchange reserves declined steeply from $16.3 billion in September 1984 to $11.3 billion in March

1985 (do Rosario, 1985). In response, the central government reimposed stricter central controls on foreign exchange and set limitations on the technology to be transferred, assigning priority to import substitution for needed technology and to projects that would promote exports. China's foreign exchange reserves have since recovered; by the end of 1988, they had reached $18 billion (New China News Agency, 1989a). China also incurred substantial trade deficits in 1985 and 1986, with a peak value of $14.9 billion for 1985; in 1987 the deficit was substantially reduced, but it increased again in 1988, although it stayed far below the levels of 1985–1986 (see Table 2.1).

Paralleling the growth of China's foreign trade, substantial changes took place in the composition of the country's exports over the years. In 1980, only 49 percent of exports were manufactured commodities; the balance comprised primary commodities. By 1987, the share of finished industrial product exports had grown to 65.5 percent, thereby reducing primary products to 34.5 percent. Textiles still accounted for about 25 percent of all exports, but finished products (ready-to-wear garments and other value-added goods) were a growing component (Frankenstein, 1988).

China's international arms trade, a component of its foreign trade, expanded rapidly during the 1980s (Frankenstein, 1988); from 1983 to 1986 arms exports increased by 167 percent over the preceding three-year period. As a result, China has emerged as the fourth largest supplier of arms to the Third World, with sales estimated at $5.3 billion during the years 1983–1986. This performance is still substantially below sales by the leading suppliers for the same period; the USSR exported $60 billion in arms; the United States, $25.5 billion; and France, $16.5 billion (Frankenstein, 1988). China has also imported military technology from the United States (avionics, munitions technology, naval engines), from the U.K. and Italy (tank and naval technology), from Canada (artillery technology), and from France (missile technology). It has become clear that China has pursued purely economic goals in its arms trade and has largely ignored political interests. For example, intermediate-range ballistic missiles were sold to Saudi Arabia, even though the latter maintains an embassy in Taipei rather than in Beijing.

The bulk of China's imports have been producer goods rather than consumer goods, underlining the fact that China is not a consumerist economy, largely because of the low per capita income. In any case, it is not the aim of the government to develop consumerism (Frankenstein, 1988). According to this author, China must therefore rely on export-driven incentives to develop technology. The economy is not considered to be export-led at present, since only 6 percent of the labor force is employed by exporting industries accounting for 14 to 15 percent of total national export. This compares with 55 percent of the labor force employed in export industries in Taiwan and 37 percent in Korea (Frankenstein, 1988). On the other hand, the considerable domestic demand for industrial equipment could also contribute to innovation.

The capitalist countries of the industrialized world have been China's major trading partners. For the years 1979–1986, Japan has had a 24.9 percent share of China's average annual two-way trade; EEC countries (including FRG), 22.6 percent; Hong Kong, 21.5 percent; and the United States, 11.8 percent. All analysts agree that China is committed to participate as fully as possible in the world capitalist economy (Frankenstein, 1988), since this includes the largest markets for China's exports and at the same time the best potential sources for much needed technologies and investment capital. However, since 1979, China has also developed a number of new trade partners; this development has important political implications (Frankenstein, 1988). Trade with the USSR predominated during the 1950s and subsequently decreased to a low point during the Cultural Revolution. However, between 1979 and 1986, trade between the two countries increased from $492 million (2 percent of Chinese foreign trade) to $2.7 billion (4 percent), while Eastern Europe's share of foreign trade decreased from a high of 7 percent to 2.9 percent over the same period. It follows that the increase in trade with the USSR occurred at the expense of China's trade with Eastern Europe and the Third World, rather than at the expense of trade with the industrialized capitalist countries.

The increase in trade with the USSR paralleled the relaxation of political tensions between the two countries, as might be expected. For China, USSR is a convenient market for industrial commodities that cannot compete in the more sophisticated markets of Japan, Western Europe, and the United States. Likewise, USSR is a good source of cheaper, if somewhat inferior, goods. Also, trade by barter, co-production, and buy-back arrangements fulfill China's aim of economizing on foreign exchange (Frankenstein, 1988). China's trade with South Korea and Taiwan, although still largely by indirect channels, has also expanded. This development also indicates China's decision to join in the capitalist market and the free world economy (Yun, 1989; Clough, 1989).

Frankenstein (1988) has pointed out that China's economic foreign relations policy still faces a number of potentially serious problems. For one, total Chinese foreign trade, in spite of its spectacular expansion (see Table 2.1), is still small compared with that of the United States or Japan, which, in 1987, totaled $677.0 billion and $375.1 billion, respectively (World Bank, 1987). In 1987, it was also still somewhat smaller than the foreign trade of the much less populous newly industrialized countries (NICs): South Korea's and Taiwan's foreign trade totals amounted to $88.0 billion and $83.4 billion, respectively.

HISTORY OF CHINESE TECHNOLOGY ACQUISITIONS

The history of technology transfer to the People's Republic of China between the early 1950s and about 1982 has been called the *four waves* (Ho and Huenemann, 1984). Table 2.3, reproduced from this reference, lists the value of imports of machinery and equipment from 1952 to 1982. Four distinct periods emerge:

TABLE 2.3 China's Imports of Machinery and Equipment: 1952–1982.

Millions U.S. $			
Year	Imports	Year	Imports
1952	$193	1968	$ 235
1953	276	1969	214
1954	381	1970	398
1955	411	1971	481
1956	545	1972	524
1957	566	1973	797
1958	715	1974	1,605
1959	933	1975	2,013
1960	840	1976	1,716
1961	272	1977	1,171
1962	102	1978	2,033
1963	100	1979	3,832
1964	162	1980	5,352
1965	302	1981	4,661
1966	443	1982	3,401
1967	335		

Source: Ho and Huenemann, 1984. Data for 1952 to 1974 are from U.S. government sources quoted in A. Doak Barnett, *China's Economy in Global Perspective* (Washington, D.C.: The Brookings Institution, 1981), p. 190.

Data for 1975 through 1982 are from C.I.A., *China: International Trade Annual Statistical Supplement,* February 1982, p. 53; and *China: International Trade, Fourth Quarter, 1982*, June 1983, p. 11.

1. *1952–1960.* During this period China imported 256 complete plants from the USSR and a similar number from East European countries. This undertaking represented the technology core of the First Five-Year Plan and has been called the "most comprehensive technology transfer in modern industrial history" (Ho and Huenemann, 1984). About 50 to 70 percent of the equipment was imported from the USSR, accompanied by full technical documentation and specifications in order to facilitate installation and maintenance. Also, the availability of complete sets of blueprints made possible the production of spare parts and even the duplication of entire plants, thus encouraging local development of production capability. During this period, it is estimated that about 10,000 Soviet engineers and technicians worked in China and that some 15,000 Chinese technicians were sent for training to the USSR. The period of close technological cooperation with the USSR came to an abrupt end in 1960 when Stalin withdrew all Soviet citizens from China after the rupture of political relations between the two countries. The majority of the projects had been completed by then, and the situation was ameliorated by the fact that the East Euro-

peans only partly withdrew. Nevertheless, Chinese attitudes in the 1980s concerning technology transfer bear the effects of China's experience with the USSR.

2. *1962-1967.* China soon sought replacements for the lost relationship with the USSR and turned to Japan and Western Europe. According to various sources, between 46 and more than 80 complete plants (in large part, turn-key projects) were imported, including two synthetic textile plants from Japan and a truck factory from Europe. This "wave" of equipment imports was ended by the Cultural Revolution, which began in 1966 and effectively disrupted all industrial activity. For example, work at the truck factory located at Wuhan was begun in 1964, but the first vehicle was only completed in 1977.

3. *1970-1977.* The National Technical Import Corporation (Techimport), one of the foreign trade corporations under the Ministry of Foreign Trade, was revived in 1972. In 1973, the State Council appropriated the equivalent of U.S. $4.3 billion for the import of equipment over a four-year period (the "Four Three Program"). By the end of 1977, contracts in the amount of $3.5 billion had actually been signed. Imported plants included the Wuhan Iron and Steel Works and the Wuhan Rolling Mill, whose productivities became the subject of controversy and debate. The technology was imported from Japan and West Germany. A significant number of petrochemical plants were acquired as turnkey projects during this period, including 11 ammonia plants from Kellogg Corporation (1973-1974) and four ethylene plants from Combustion Engineering Lummus Corporation (at Yanshan in 1983 and at Nanjing, Shengli oil field in Shandong, and Jinshan in 1978) (Weil, 1983).

4. *1977-1982.* China signed foreign contracts worth $6.4 billion in 1978. As already mentioned, some of these were postponed or even canceled after the policy of readjustment was enacted at the Third Plenum of the Eleventh Central Committee in December 1978 and a general reorientation from emphasis on turnkey projects to technology acquisition was announced by Party Secretary Zhao Ziyang in his speech to the Fourth Plenary Session of the Fifth NPC in late 1981.

Early Technology Acquisitions

A number of problems were found to have limited the effectiveness of the technology acquisition program during the third wave ("Four Three Program"). These problems were debated by Chinese analysts from 1979 to 1981 (Ho and Huenemann, 1984, pp. 15-20). The major points made were as follows:

1. Delays in construction schedules by one to three years were registered in 11 out of 24 cases examined.

2. Many of the new plants were operating substantially below capacity, in some cases below 50 percent of capacity (four ammonia plants); and levels of 60 to 75 percent of capacity were widespread.
3. The return on investment was judged to be poor. Three-quarters of the projects failed to recapture the investment within three to four years, which was considered to be a reasonable period. For example, the Wuhan Rolling Mill was cited as being particularly poorly utilized, operating at 20 to 30 percent of capacity. Production only began after a number of problems had been solved. The product turned out to be of high quality, but its price was not competitive on the world market.

Other criticism was leveled at the Wuhan Iron and Steel Works, but its defenders pointed out that a number of important technological advances had been introduced into China in the course of this project and a significant amount of profits and taxes had been earned by the state from the Works' operation. Also, Chinese ability to absorb advanced technology had been demonstrated, even if it had taken longer and had cost more than necessary.

A Chinese analyst, Lin Senmu, writing in a Chinese economic journal in 1981, analyzed the mistakes made in 1978 and listed four lessons to be learned from the experience. The Baoshan General Iron and Steel Works in Shanghai, whose construction was begun in 1978, was the most prominent example of a project fraught with problems. It became the subject of extensive studies.

1. The domestic inputs required by an imported factory must be studied in advance and their availability must be assessed. The inputs included fuel, utilities, housing, and transport. It was found that 22 key projects of the "fourth wave" required unmanageable quantities of petroleum (more than 10 megatons) and coal (20 megatons).
2. Local resources including domestically produced components should be used as much as possible and excessive automation should be avoided.
3. Technical and economic feasibility studies should be required before approving a project.
4. China must advance its knowledge about the Western business world.

As seen from Table 2.1, the growth of Chinese foreign trade was substantial during the 1970s and early 1980s, but control over commodities and capital remained tight. In early 1982, the State Council restricted imports in order to protect domestic development. The Chinese government continued its stress on technology, but the principal aim now was to *absorb* technology, including management and marketing skills, from abroad. Previously, it has been thought sufficient to import a prototype of a given product (instrument, machine) and then to attempt to produce it by employing a process often referred to as *retroengineering*; now it was concluded to be more effective to obtain the entire system and the processing technology concerned. Thus China began establishing trade and technology agreements such as licensing agreements and joint ventures, the latter being considered the most time and cost effective by the Chinese authorities. Cooperative management of joint ventures offered the best opportunity for exposing Chinese managers to foreign management methods and for improving their management expertise (Simon, 1987). The aim of foreign exchange conservation would also be achieved by such cooperative ventures. Purchases of equipment and components from the technology supplier were usually essential for the success of the project and served as inducements to the foreign affiliate.

The renovation of existing industrial facilities, as opposed to the construction of more plants, was advocated about 1982, but an increased emphasis on this policy was included in the Seventh Five-Year Plan (1986–1990) (Simon, 1987). The total investment in technical transformation and equipment renewal for state-owned enterprises and institutions was set at 276 billion yuan ($74.6 billion), an 87 percent increase over the previous five-year plan. About 600 major modernization projects were defined, with the machine building and electronics sectors receiving first priority. The aims of renovation included improvement of economic performance, enhancement of product quality, energy conservation, reduction of raw material consumption, development of new products, improvement of occupational safety, and reduction of environmental pollution (Simon, 1987).

According to Chinese analyses, the impact of technology transfer projects was severely limited by a large number of difficulties (Simon, 1987); four of the most serious are listed below.

1. The activities of responsible Chinese organizations were uncoordinated. They were competing against one another.
2. In the course of decentralization, Beijing transferred some decision-making authority to the provinces, but limitations on such authority often prevented success of negotiations. On the other hand, local agencies often deviated further than desirable from central guidelines to assure success.
3. There has been excessive duplication of imported technology and equipment. For example, it has been estimated that over 100 color television production lines have been imported. The resulting capacity surpasses demand by a wide margin. The excess product could not be exported because of limited quality control and product design difficulties.
4. Not enough attention has been paid to technology assimilation capability. The central authorities concentrated attention on export control limitations imposed by the industrialized countries and succeeded in having these controls significantly relaxed since about 1984. However, the present limitation is determined by China's technical and administrative absorption capability.

Recent Technology Acquisitions

In view of these findings, two pieces of legislation were introduced in May and August of 1985, instituting new procedures of review and evaluation before any technology transfer from abroad to China could be approved (Simon, 1987). The new regulations included the following provisions:

☐ Restrictions on the use of transferred technology were prohibited.

☐ The level of appropriateness of the technology was required to meet one of eight criteria, which measured its contribution to the economic modernization of the country.

☐ The foreign supplier had to guarantee that the technology was free from defects and that it would allow the technology recipient to achieve the stated objectives.

☐ The term of an agreement was limited to ten years.

☐ MOFERT or its designated representative had to approve the contract.

Some of these provisions lacked precise definition and left a great deal to the discretion of the Chinese authorities. To what extent the supplier would be held responsible for the quality and performance of the technology remained unclear, since the foreign participant's control over environment, inputs, and product quality was quite limited. Nevertheless, the Chinese government viewed the regulations as helpful to resolve inconsistencies. It was also clear from the enacted provisions that the Chinese authorities were disappointed with the technological progress made as a result of foreign participation in the Chinese economy (Simon, 1987). For instance, in-depth training of Chinese personnel was not common to all cooperative manufacturing projects, although some fulfilled Chinese expectations. The new regulations, in addition to raising the level of approved technology import projects, also increased the bargaining power of the Chinese side because of the greater mandated selectivity. As a result, the new provisions also had the potential for increased deterrence of foreigners from giving support to China's aims at modernization.

A breakdown of technology transfer project statistics for the year 1986 is instructive (China Encyclopedia Yearbook, 1987). In 1986 MOFERT approved 744 contracts for a total value of $4.46 billion, an increase of 50.7 percent over the previous year's total value (the number of contracts increased by 10.9 percent). Of this total, $2.9 billion concerned energy projects, including the large nuclear power plant imported from France under construction in Daya Bay, Guangdong Province near Hong Kong, and ten complete fossil fuel electric power plants with total capacity of 5,600 megawatts (the nuclear power plant accounted for 1,800 megawatts). By country, the breakdown of the technology imports during 1986 was as follows: United States, $657 million or 14.7 percent; EEC countries, $2.13 billion or 47.9 percent; USSR and Eastern European countries, $810 million or 18.2 percent; Japan, $787 million or 17.7 percent. Of the total, $800 million or 18 percent accounted for contracts containing as principal component licensing, consulting, servicing, and co-production. Contracts for equipment import totaled $3.85 billion or 86.4 percent. It should be noted that the last two categories together exceeded 100 percent of the total but details of the overlapping categories were not reported. Also, the statistics cited here do not include technology transfer values as part of foreign-invested projects, such as equity joint ventures or contractual joint ventures. It should also be noted that the figures quoted here are not necessarily comparable to those listed in Table 2.3, since the categories included in the statistics are not unambiguously defined in the literature. Nevertheless, the value of equipment imported in 1986 ($3.85 billion) is of similar magnitude to the machinery and equipment imports listed in Table 2.3 for the years 1978-1982.

FOREIGN INVESTMENT

Investment Volume

So far, foreign investment volume has fallen short of China's expectations. Most foreign-invested enterprises are small or medium-sized projects, with Hong Kong being the leading source of capital by far, followed by the United States. Although Japan has been the leading trade partner of China, Japanese companies have not invested heavily, since industrial leaders have historically not considered China to be an attractive investment site (Frankenstein, 1988). However, since 1986 a number of factors have forced a change in policy (Simon, 1987). Three factors stand out: (1) The export of final products has been impeded by China's crunch on foreign exchange and (2) by the appreciation of the yen. Also, (3) China has exerted increasing pressure on Japanese companies to invest. The final evidence for rising Japanese investment in China is not yet in, but some observers have reported encouraging signs.

Total investment contracts approved by the Chinese government between 1979 and August 1988 were reported to be worth $25.6 billion, but only $10 billion have actually been utilized. The approved amount included off-shore oil exploration projects with a total value of $2.9 billion. The data above apply to 13,000 approved foreign-invested enterprises, 6,000 of which are actually in operation. Of those in operation, 1,500 have been reported to have achieved either balance or a surplus of foreign exchange (New China News Agency, 1988c). Investment commitments by U.S. companies in China for the period 1979-1986 amounted to $2.6 billion, of which about $1 billion was in off-shore oil exploration projects. Table 2.4 gives further details of these statistics.

TABLE 2.4   United States Direct Investment in China

	1979–1985	1986	Total
Equity joint ventures			
Number of contracts	138	81	219
Dollars committed (billions)	$0.265	$0.263	$0.528
Contractual enterprises			
Number of contracts	38	15	53
Dollars committed (billions)	$0.890	$0.180	$1.070
Wholly-owned enterprises			
Number of contracts	5	2	7
Dollars committed (billions)	$0.005	$0.001	$0.006
Off-shore oil exploration			
Number of contracts	21	4	25
Dollars committed (billions)	$0.960	$0.068	$1.028
Totals			
Number of contracts	202	102	
Dollars committed (billions)	$2.120	$0.512	$2.632

Source: Sullivan, 1988.

Chinese Policies and Legislation

About 200 pieces of legislation relevant to foreign investment and trade have been promulgated since 1979, the most significant being those regulating contracts and taxes (Lubman, 1986; Cohen, 1988). The patent law, which was introduced in April 1985 (Simon, 1987), has been of special interest to foreign companies, since concern for the protection of intellectual property has been widespread among foreign investors and licensors. Although questions remain concerning implementation of these laws, their mere enactment signals China's commitment to creating an infrastructure compatible with international commerce and trade (Frankenstein, 1988).

The Joint Venture Law passed by the Seventh NPC in March 1987 embodied reforms introduced in the course of the years. Beginning from February 1986, the State Council extended the maximum term of most joint ventures from 30 to 50 years, and even longer terms were considered in special cases. Also new measures were instituted to assist joint ventures and cooperative joint ventures in satisfying their foreign exchange requirements; for example, Chinese government agencies with foreign exchange holdings would be allowed to pay for products in whole or in part with foreign currency. In addition, government departments responsible for joint ventures were empowered to use foreign exchange surpluses of other joint ventures to help balance shortfalls (New China News Agency, 1986). More recent regulations instituted Foreign Exchange Adjustment Centers in several cities where foreign-invested joint ventures and cooperative ventures could buy and sell foreign exchange among themselves, subject to confirmation by Chinese authorities.[2] Rates have varied according to supply and demand, but have remained mostly in the range of 6.50 to 7.00 yuan to the U.S. dollar (Yowell, 1988).

According to the strategy developed by the Chinese government, the capital required for the modernization of the country's industry was to be obtained in part from the export of energy resources including oil and coal as well as certain nonferrous metals in world demand such as tungsten, vanadium, and tin. Foreign investment in the development of energy resources was also sought; many foreign companies made substantial investments in exploration of offshore oil and gas resources, the total approved investment amounting to $2.9 billion by August 1988, with U.S. companies having committed about one-third of the total (see Table 2.4 and related discussion). Occidental Petroleum Corporation holds a 25 percent share in a $700 million joint venture for the development of the open-pit Pingshuo Coal Mine in Shanxi Province. The development of energy resources was recognized as having high priority not only as providing valuable export commodities but also to remedy the growing shortage of energy supply to Chinese industry.

Sectoral orientation of investment has been of concern to the Chinese leadership. The Seventh Five-Year Plan (for the years 1986–1990) listed as priorities for development and modernization the sectors of energy, computers, electronics, transport, communications, food processing, and building construction. Although China has obtained concessionary loans from Japan and other countries, it has failed to attract sufficient investment in these sectors (Simon, 1987). Most investments during the years 1979–1984 had gone into light industry, real estate, and tourism. Hotels, apartments, and office buildings are attractive projects because they earn large surpluses of foreign exchange and therefore are free from the most troublesome problem encountered by investors.

Foreign Company Experiences and Concerns

Foreign companies with investments in China have long complained about a number of problems they have encountered, and it has been assumed that these hurdles to profitable business activity have kept foreign investment below levels targeted by the Chinese government. A number of these, which still persist, have been discussed by Frankenstein (1988). Bureaucratic and operational difficulties discourage investors. Services available to businesses are both expensive and of low quality, making China an expensive place for business operation. It is difficult to recruit and retain skilled staff. Intervention by party officials is often arbitrary and unpredictable. The bureaucracy remains impenetrable to outsiders. Negotiations still require a relatively long time and are often unproductive. Access to the domestic market is hindered by great bar-

---

[2]Enterprises in the tourist sector, viz. hotels, office buildings, and so forth have foreign exchange surpluses they can sell to companies short of hard currency.

riers. Foreign business personnel have reported instances of corruption, bribery, theft, and misuse of resources (in Chinese media terms, *economic crimes*) on the part of officials.

In 1986, complaints by foreign companies already invested or considering investing in China reached a crisis point after the much publicized halt in production at the American Motors Corporation's joint venture jeep production plant in Beijing in April 1986 (Simon, 1987). AMC complained about the stifling of workers' productivity and of the plant's operation by complex regulations in addition to the company's major problem, namely, securing sufficient foreign exchange. Many technology transfer projects depend, in part, on imported parts and components because of the difficulties encountered in the local sourcing of high-quality materials and parts. Also, many of the foreign-invested enterprises and other companies importing technology (for example, partners to licensing agreements) have found it difficult or impossible to export a large enough fraction of their products to earn sufficient foreign exchange. Other foreign-invested companies besides AMC that have encountered similar problems include Hitachi of Japan (color television plant in Fujian Province), Peugeot of France (automobile assembly plant in Guangzhou), and Squibb Pharmaceuticals of the United States (drug manufacturing joint venture in Shanghai) (Simon, 1987).

The Chinese government had already begun to consider major changes beginning in April 1985, motivated by foreign companies' apparent unwillingness to invest and to transfer advanced technology, as reflected in the persistently low rate of foreign investment. In addition to the concerns discussed above, foreign companies also entertained doubts regarding the permanence of the open-door policy. The government, in October 1986, responded by promulgating the "22 Provisions to Encourage Foreign Investment," which promised more preferential tax treatments, priority access to and guaranteed prices for supplies, low-cost financing, and other support for export-oriented or technologically important projects. Foreign-invested companies were to be freed from the bureaucratic hassles from which they had suffered, and local governments were made responsible for such improvements.

The Chinese leadership had three main goals (Simon, 1987): (1) to guide foreign investment toward export orientation or to the transfer of advanced technology by the use of tax incentives and the availability of favorable loans and improved labor costs; (2) to lower land-use fees and other operating costs in order to lower production costs in order to make products more competitive in the international markets; (3) to protect the autonomy of foreign-invested companies, to combat low work efficiency, and to streamline work procedures.

China hoped by means of the "22 Provisions" to stimulate foreign investment, which had not lived up to expectations and had, in fact, declined from the peak year 1985, when over 3,000 contracts worth a total of $6.3 billion had been approved. The number of contracts and their total values for 1986 were

1,500 and \$3.6 billion, respectively, and for 1987 they were 2,230 and again \$3.6 billion, respectively (Frankenstein, 1988). However, there was evidence that investment was up in 1988. The evidence comes from a comparison between investment totals for the periods 1979–December 1987 (New China News Agency, 1988b) and 1979–August 1988 (New China News Agency, 1988c). These figures indicate that \$3.6 billion had been invested during the first eight months of 1988. The provisions apply to two kinds of enterprises: (a) productive companies, which mainly produced for export and which were in a position to earn foreign exchange surpluses, and (b) enterprises where the foreign investor supplied advanced technology to be used to develop new products or to upgrade or replace old products, or enterprises producing needed import substitution products (Simon, 1987).

Response from abroad to the "22 Provisions" was mixed (Simon, 1987). Companies were encouraged by China's evident desire to be more accommodating and its continuing commitment to seeking foreign participation in the country's industrial modernization. However, other factors inside and outside China continued to impede progress (Simon, 1987). Internally, China's bureaucracy remained largely resistant to efforts at streamlining procedures. Uncertainties persisted with respect to the implementation of laws and regulations relating to contracts, taxes, and patents. Foreign exchange shortages continued to pose major difficulties in spite of new government regulations attacking the problem. The internal markets were limited and also remain protected against penetration by foreign-invested enterprises. Enterprise management and human resource management, in particular recruitment and dismissal of employees, was expected to remain problematic in spite of bona fide attempts at finding solutions.

Other factors impeding progress were external to China and beyond its control. As manufacturing technology underwent basic changes worldwide toward greater automation, China's competitive advantage based on cheap labor was questionable, at least in some industrial sectors. At the same time, other newly industrialized countries (NICs) such as South Korea, Taiwan, Hong Kong, and Singapore, were strong competitors for investment capital (Simon, 1987).

In spite of major hurdles, there have been some breakthroughs in foreign investment and some penetration of the Chinese domestic market. A number of U.S. companies have recently concluded joint venture agreements for the production of consumer goods after China decided to encourage foreigners to hold majority interests in joint ventures or establish wholly-owned companies in its quest to boost foreign investment. Procter and Gamble concluded a \$10 million joint venture agreement to produce laundry and personal care products in Guangzhou. Bausch and Lomb will produce contact lenses, a product which is likely to remain a luxury item for select Chinese consumers. After four years of negotiations, Johnson & Johnson announced a \$5 million project to produce Band-Aids in Shanghai. Seagram Co. concluded an agreement worth \$6 million after two years of negotiations for the manufacture of whiskey, sparkling wines, and wine coolers. RJR Nabisco, after six years of talks, announced a $9 million venture to make Ritz crackers, and Leaf Inc. of Bannockburn, Illinois, will produce chewing gum (Lee, 1988). These new joint ventures are encouraging to foreign investment prospects, although the ventures are all relatively small scale. The new ventures will reportedly be permitted, at least in part, to target the domestic market, which makes their approval significant. It should be remembered that 3M's (Minnesota Mining and Manufacturing) wholly-owned subsidiary in Shanghai also targeted the domestic market when it was established in 1984 (Marshall, 1987).

In spite of progress furthered by administrative regulation, problems at the enterprise level have continued. Expectations on both sides have been inflated, according to Simon (1987), and managers have continued to neglect issues of technology absorption since they have been busy coping with three types of fears: (1) they fear that their enterprise will design and produce a product that is inferior to the original import; (2) they fear the reduction of the enterprise's ability to repay loans due to lower production capacity brought about by the distraction accompanying the introduction of new technology; (3) they fear the loss of their competitive edge caused by the diffusion of the technology to other enterprises after the original aims have been realized (Simon, 1987).

According to Simon (1987), business opportunities in China are promising. However, full development of foreign trade ties and successful technology transfer is still limited by insufficient experience on both sides. China's policy to open up to foreign economic participation is only 10 years old, compared with the 20-year-old trade relations of Taiwan and South Korea. The tendency of Chinese authorities during the past few years to waver between decentralization and recentralization of the economy shows that the macro control mechanisms are not yet in place and the leadership is not ready to move forward decisively. In addition, the bureaucracy resists giving up central control, and they also view the repatriation of profits as a concession and wish to minimize its implementation. Nevertheless, the recognition is steadily growing that foreign companies must be allowed to make and repatriate profits if they are to find investment in China attractive.

The new and more restrictive regulations imposed on technology transfer may result in deterring foreigners but they may also serve to clarify which type of project is considered desirable by China. If projects supporting exports or producing substitute local goods for needed imports (import substitution) are indeed given greater attention and enjoy smoother sailing, the technology import regulations of 1985 may achieve a mutually beneficial result after all (Simon, 1987). Measures for implementation of such preferentially supportive policy have recently been announced and will be enacted during 1988–1989 (New China News Agency, 1987b). Also, China has defined a list of sectors in which priority treatment will be accorded to foreign-invested enterprises (China Economic Indicators Database, 1988).

FUTURE OUTLOOK

The outlook for foreign trade and technology transfer for the immediate future is uncertain.

The political events of May-June 1989 will undoubtedly have far-reaching effects on the short term future foreign trade relationships of China. In the first instance, the Western industrialized countries, viz. the U. S. and EEC have, at least temporarily, taken steps which amount to economic sanctions. The U.S. has discontinued the sale of military equipment and has limited the export of technology. In addition, President Bush has announced that the U.S. will withdraw its support for World Bank loans to China and the major EEC countries have concurred with this position. The loans from the World Bank have been of great importance to China since they have supported educational and agricultural development at concessionary interest rates, averaging at about 4%.

Japan's reaction has been more low-key although it joined in the declaration by the seven industrialized countries meeting in Paris on July 16, 1989, which called on China to desist from its suppression of the democracy movement and from on-going human rights violations. The USSR has only expressed mild regrets over the repressive measures employed by the Chinese government. The differences between the reactions of the United States and the EEC countries on one hand and Japan and the USSR on the other can be correlated with the cultural differences and the varying levels of concern regarding human rights embedded in the traditions of these countries. The USSR may actually be poised to reap advantage from the situation as China may look for more technology transfer from Eastern Europe and Russia. There are clear advantages for China in trading with the socialist countries since their markets are less demanding in terms of product quality and financial terms would be easier for China. Japan may also continue to advance its trade position in China during a period of lower level activity on the part of the United States and the EEC countries.

Clearly, all foreign analysts must conclude at this time that the political climate in China has an inherent instability, which will prevail for an extended period, perhaps for 5-10 years, or even longer. Even as the aging conservative leaders of China will depart from the scene in due course, there is no clear leadership in sight which could steer the country onto a stable path of progress and development. It is, therefore, to be expected that China will once again experience a period characterized by conflict and struggle for succession.

Clear evidence of widespread discontent among the workers who joined the student protests has emerged and the government has responded with great severity to workers involved in the demonstrations of May-June 1989 while taking clearly less severe measures against students. It may be expected that the discontent and subsequent cruel suppression will result in passive, or in some instances in overt, resistance by industrial workers as well as engineers. As a consequence productivity will suffer over the near future.

Trade and business relations between the United States and China may be expected to resume after a cooling off period. This tendency will probably be expedited by the expected growing business activity between China and the USSR and perhaps also with Japan. However, foreign companies will have to adapt to the new conditions of political and social instability, prevailing in China. The reality of expected lack of political stability will mandate a general reluctance by American and other foreign companies to invest in equity joint ventures. Also Japan and Russia are likely to be wary of increasing investments at this time. The appropriate business strategy will therefore favor sales and eventually also technology transfer governed by licensing agreements, counter trade, compensation trade and other non-equity arrangements.

CASE 1

Shanghai-Foxboro Company Limited: The U.S. Perspective

INTRODUCTION

In 1986, the Shanghai-Foxboro Company Limited (SFCL) was one of the oldest technology transfer joint ventures between the United States and the People's Republic of China (PRC). During President Reagan's visit to China in April 1984 to promote bilateral trade, SFCL was the plant chosen for his visit. As the press coverage noted at the time, SFCL was an example of how cooperative ventures could succeed even across vastly different cultures.

Entered into by the Shanghai Instrumentation Industry Company (SIIC) of the PRC and the Foxboro Company of the United States, a 20-year joint venture agreement was signed in Beijing on April 12, 1982, making SFCL the first U.S.-China joint venture involving the transfer of high technology.[1] The new company, Shanghai-Foxboro Company Limited (SFCL), displayed its first instrumentation control system, assembled in Shanghai, at the Instrument Exhibition in Beijing in April 1983 and made its first customer shipment in June of that year. Between 1983 and 1986, SFCL produced electronic analog control devices, resonant wire transmitters, and digital computer systems devices

---

[1]Shanghai Instrumentation Industry Company was subsequently renamed Shanghai Instrument Company (SIC). By 1987, SIC, which was a conglomeration of several production facilities primarily in the instrumentation industry, had been dissolved as part of the government's drive toward decentralization. More autonomy was given each production facility.

and conducted over 125,000 man-hours of training. (Exhibit 1 on page 76 summarizes the major events in SFCL's history.)

This case describes the role of SFCL in Foxboro's global strategy and the process by which it managed the negotiations, agreements, start-up, and ongoing management issues raised by the joint venture in China. The case is based on a series of interviews with senior executives of the Foxboro Company and has been prepared with their cooperation. The research was supported by a grant from the U.S. Department of Education.

Foxboro Company

The Foxboro Company of Foxboro, Massachusetts, was founded in 1908. By 1985, it was a leading worldwide supplier of pneumatic and electronic control instruments and computer control systems of varying size and complexity. Its primary customers were the major process industries including chemicals, oil, electric utilities, water and waste treatment, and so forth. Foxboro's products were used to measure and control such process variables as flow, temperature, pressure, and liquid level, as well as composition of the materials being processed. Products were sold individually or engineered into process management and control systems for specific customer situations. Although process-control instruments usually represented a relatively small portion of the total investment in these basic industries, they significantly influenced the output and quality of the total process.

In computing revenues, Foxboro grouped its products into four categories: (1) electronics systems and instruments, (2) pneumatic instruments, (3) parts and accessories, and (4) panels and services.[2] Electronics systems and instruments were the company's newest product offerings and generated 31 percent of 1984 orders. Pneumatic instruments generated 19 percent. Although pneumatic instruments represented a mature product offering, they remained the preferred products for many installations and provided ongoing sales opportunities for the maintenance of a substantial installed base.

In 1984, Foxboro's sales exceeded $500 million. It employed 10,250 people and operated 23 plants in 10 countries. In addition, it had approximately 180 sales and support centers in 100 countries. Only half of Foxboro's sales came from within the United States. It derived 16 percent of its volume from Western Europe, 14 percent from the Middle East and Africa, and the remaining 20 percent from the rest of the world. From an industry standpoint, Foxboro's three largest customer groups were chemical processing (24 percent), oil and gas (29 percent), and pulp and paper (11 percent). (Exhibit 2 on page 78 summarizes Foxboro's financial data.)

Foxboro had the broadest product line and the largest share of the available world market for its process-control equipment. In addition, it had high visi-

---

[2]Foxboro Company, *Annual Report,* 1984.

bility. In nearly every category of instruments or systems listed in an annual recognition survey conducted by *Chemical Engineering* magazine, Foxboro was ranked number one. And in most cases, Foxboro had retained this ranking since 1970. Foxboro's strategy was distinguished by its focus on process management and control; by its status as a worldwide service organization that maintains its commitment to total customer support; and by its extensive research, development, and engineering programs. Foxboro consistently spent 7 to 8 percent of its revenues on R&D and employed almost 700 professionals in its technology organization.

In the context of being a worldwide leader in instrumentation and control, Foxboro's executives viewed China as one of the last, large undeveloped markets in the world for oil refining plants, petrochemical plants, and other process industries that were Foxboro's primary customers. In industrially advanced countries, these process industries had reached mature stages of their life cycle.

DEVELOPMENT

Early Contacts and Approaches

Foxboro had done business with Chinese industry in the 1930s and 1940s until the People's Republic suspended relations between the United States and China. The trade window did not open again until President Nixon's trip to China in 1972. Even then, U.S. government export controls severely constrained the kinds of products and information that Foxboro could send to China; therefore, Foxboro started to establish business relations in China through its West European subsidiaries.

Early in 1975, for example, the British Board of Trade sponsored an exhibition of British-manufactured goods in China. Foxboro participated through its British subsidiary and exhibited the mechanical and pneumatic products produced by that subsidiary. The exhibits were staffed with ethnic Chinese from Foxboro's Singapore office.

From 1975 to 1978, whenever there was a nationally or privately sponsored exhibit in China that was relevant to Foxboro's products, Foxboro participated. Mr. Gerald Gleason, vice president of Foxboro, described the impact as follows:

> We were using Chinese staff from our Singapore office in these exhibits, and they began to make contact with users in China who would invite them to visit their plants. Our people would be given a hotel room, most technical people around would know they were there, and our people would end up working all day and night because there was such a pent-up demand for up-to-date technology in our field. With these visits came orders. The orders would be entered in Singapore,

and we'd go through the usual export control routine in this country for filling a Singapore order.

Toward the end of 1978, the National Bureau of Instrumentation and Automation contacted our Far East manager and asked, "Would your management in the United States be interested in more of a relationship than this sales relationship?[3] They checked back here, and we said, "Well, we don't know. But we would certainly be open to discussing it." As a result of which, in late 1978, we had an invitation to send a delegation to China.

In February 1979, Foxboro sent a six-member delegation to China for three weeks. The delegation from Massachusetts consisted of the company chairman and CEO, the president and COO, the treasurer, and Mr. Gerald Gleason, vice president. They were joined by Foxboro's Far East general manager and the Singapore general manager.

The delegation was hosted in China by the head of the National Bureau of Instrumentation and Automation and taken to tour factories that were typical of the Chinese instrumentation industry and also to user industries such as oil refineries and chemical plants.[3] The delegation also met representatives from research institutes, universities, the Foreign Investment Commission and officials in Beijing, as well as provincial authorities in Shanghai and Guangdong.[4] At the end of three weeks, the senior officials in Beijing asked Foxboro's top management, "Well, what do you think?"

The response of Foxboro's executives was that there seemed to be a long road ahead in terms of developing a modern Foxboro-type instrumentation plant in China, but the possibilities looked interesting enough for Foxboro to explore the opportunity. At that time, Foxboro's CEO asked Gerald Gleason, Foxboro's vice president and member of the delegation, to be the primary contact with the Chinese side. The Chinese officials, in turn, appointed a deputy director of the Bureau as their contact. On Foxboro's side, Gleason handled the negotiations up to the point of signing the agreements. His assignment had no special title associated with it. He reported directly to the President's office, which included Foxboro's chairman and president.

Foxboro did not have a separate international division. (Exhibit 3 on page 79 shows Foxboro's organizational chart.) Foxboro was run on a matrix structure with two dimensions: (1) corporate discipline skills, for example, manufacturing, technology, sales, and so forth, with international responsibility for each discipline; and (2) geographic areas, each of which reported to an area general manager responsible to the president and chief operating officer (COO).

---

[3]The National Bureau of Instrumentation and Automation was a bureau of the First Ministry of Machine Building, the ministry most operative in Foxboro's field. The ministry was subsequently renamed Ministry of Machine Building Industry and changed again to the State Commission of Machinery Industries.

[4]The Foreign Investment Commission has since been absorbed into MOFERT (Ministry of Foreign Economic Relations and Trade).

Information on Chinese Markets

Obtaining data to estimate market size and feasibility in the early stages was difficult. Very little published information was available, and field market research interviews were not a commonly acceptable means of collecting data. Foxboro's executives, therefore, relied on "informed gut feelings." They knew, in a general sense, how many oil refineries and chemical plants existed in China, but they knew very little about their capacity. So they made broad assumptions about average plant output, volume, and so forth, and ran a series of "what if" computer "games" to estimate a capacity for their venture that would best balance the constraints of market size, foreign exchange, and manufacturing feasibility. Foxboro recommended, and the Chinese side accepted, the proposed capacity and magnitude of the venture.

Choice of Partner

The Chinese government proposed a specific joint venture partner for Foxboro. The company was very pleased with the choice and did not evaluate alternative collaborators. As one Foxboro executive remarked,

> The Chinese government picked a fine one for us in Shanghai. Probably 40 to 60 percent of all the good electronics being built in China right now is built in Shanghai, and it was our good luck that the Shanghai Telecommunications and Instrumentation Bureau is probably the most powerful operation of its kind in China.[5]
>
> The Chinese are anxious for the joint venture to be successful, so all the people we get are the absolute cream, whether they are interpreters or engineers.

In terms of the selection by the Chinese government of Foxboro over its other competitors in process control such as Honeywell, Fisher Controls, and so forth, Foxboro's executives believed that its reputation for technological leadership and professional management, as well as its known willingness and experience with transferring state-of-the-art technology to its foreign affiliates were key factors in Foxboro's selection.

NEGOTIATIONS

Negotiating Personnel

The primary negotiator of the initial agreement from Foxboro's end was Gerald Gleason. Gleason had been in international work since the 1950s and had negotiated agreements in many places, including the Soviet Union and Romania. Gleason kept Foxboro's chairman and president informed of the progress in the negotiations. In addition, he constantly checked with the Depart-

---

[5]This bureau was subsequently renamed the Shanghai Instrumentation and Electronics Bureau.

ment of Commerce and, less frequently, with the Department of State. As Gleason put it, "I didn't ask for approval of our agreements, but I said, 'If I'm saying something you don't want to hear, tell me now so I can change it.' " Gleason used the Foxboro legal department only as backup consultants in the initial stages to avoid legal or adversarial confrontations with the Chinese negotiators.

Gleason also involved Foxboro's Far East manager in most of the negotiations. This choice was particularly meaningful for two reasons: (1) The manager was located in Singapore and could make frequent trips into China to resolve minor issues that did not justify a trip by Gleason from the United States; (2) he spoke Chinese fluently and could monitor the accuracy of the translation during the negotiations, a critical point, since Gleason had decided to rely on the Chinese interpreter rather than have one of his own. Gleason believed that the Chinese were likely to be very comfortable with their own interpreters and that they would not be so at ease if a Western interpreter was included. In addition, Gleason believed his action was likely to be construed as a signal of trust by the Chinese.

Although the Chinese interpreters were technically very good in English, in the initial stages they did have some difficulty understanding Western commercial language. Gleason explained as follows:

> If you don't know the word *profit,* you can't translate it. . . . So as a backstop, I would generally have in each meeting, my Singapore manager or one of his assistants, whom we would never use as an interpreter, but only as a monitor. And if he thought that Madame Zhang had not expressed something properly, he would interrupt and say, "Madame Zhang, I don't think that is what Mr. Gleason was trying to say. May I try it for you?" And that's the way we got over the hump of her not knowing commercial language in the beginning. . . . And they never objected to his being there because his reason for being there, as I always said, was because he was our Far East manager and he, therefore, had a place at the negotiations. . . .
>
> But he was very carefully constrained about what he could say on his own. If he got carried away and started going off in Chinese to the group, I would ask him to slow down immediately."

Venue of Negotiations

To keep negotiations mutual from the start and to signal that "We're not going to keep coming to you, you have to come to us," Foxboro tried to achieve a balance between meetings in China and in the United States. Both venues had their advantages and disadvantages. The major advantage of meeting in the United States was that Foxboro could show the Chinese negotiators around the plant, give them demonstrations, and build greater credibility and assurance about the actual technology to be transferred. Negotiating in China, on the other hand, meant that the Chinese could, whenever necessary, obtain clarifications and authorizations from their superiors to expedite the negotiating process.

Foxboro's executives would usually go to China for relatively short periods and return to the United States whenever there was an impasse. The Chinese delegations, on the other hand, consisted of between six and ten people and would stay as long as six weeks. However, only part of their time was spent at the negotiating table; the rest was spent learning about Foxboro.

Gleason described some of the quirks in the negotiating process:

We got into a real hang-up one day about a *force majeure* clause. I related this to discussions I had had with the Soviet Union, which took a very negative view about such clauses. I made an incorrect "bridge" and assumed that this would be a problem with the Chinese, too, because they objected very strenuously to it. But the situation turned out to be quite simple, although it took me a long time to dig my way through it. It turned out that they were objecting to the statement in the clause that referred to "acts of God." When we changed that to "natural acts," they had no problem with it.

Another problem occurred when I insisted that there be two versions, both official, one in Chinese and one in English. They insisted that the only official version would be in Chinese and there would be a translation into English. I said, "No. I don't know Chinese. I won't sign an agreement unless I can sign an official version in a language in which I'm more fluent." It went round and round for several days. On the third day, my secretary came across a press report on the first U.S.-China trade agreements that said there would be two versions of this agreement, both official, both formal, and one would be in English and one in Chinese. These people didn't know that. They were in the United States at the time, so they could not easily pick up the phone and say, "Is that all right?" So when they came in, I asked their interpreter to read the report in Chinese to the group. Then they all smiled and said, "Ah, we have wasted two days. But now it's OK, we will have two versions."

Negotiating Styles and Tactics

While Gleason spearheaded the negotiations, Foxboro's Singapore salespeople continued the direct sales, plant visits, and participation in exhibitions as usual. Because these earlier activities were generating fairly good business, Foxboro was not in a special hurry to conclude its joint venture negotiations. Since Foxboro's work force commitment to the joint venture included only Gleason and, to some extent, its Singapore manager, continuing negotiations were also not very expensive to the company. In addition, Foxboro's executives did not expect the Chinese venture to contribute significantly to Foxboro's profitability in the short term. Hence, Foxboro's approach was to "walk slowly and get there on a sound basis, not rushing into an agreement."

Gleason's style with China balanced firmness and perseverance with a great deal of patience. He described it as follows:

You get war stories about all the Americans who go over there and say they won't go home unless they reach an agreement. Well, I never stayed that long. I would go and when I knew I'd reached an impasse, where I wasn't about to concede,

I'd go home. I never stayed there more than three weeks, sometimes shorter. I would say, "When you have a new idea on this, let me know. We'll talk some more." Sometimes four months would go by and we wouldn't be in contact with each other at all. And then, eventually, I would receive something in the mail, a new draft or something with a little note saying to try this and see whether it does anything better. And usually there was some form of concession or a new way to go around the barn. I made concessions too, but not on what I considered essential points.

Although the Chinese delegations were reluctant to discuss issues that fell outside their domain without prior guidance, they had considerable authority in negotiating operational issues in their own area. In general, Foxboro's executives were satisfied with and respected the Chinese negotiators. Edward McIntyre, Foxboro general counsel in 1982, described his view:

The negotiations and discussions were much easier than I anticipated. There was no shadow boxing. However, I should say that the Chinese were very good negotiators. There were times when someone from Foxboro would raise an issue, and they would not respond. And there would be five minutes of total silence, which is hard for Americans to take, and we'd end up volunteering answers. . . . So maybe they did have a very clever negotiation strategy, they just seemed to adopt a very fair attitude to doing what would be best for the joint venture.

In terms of the atmosphere in China, Gleason preferred it to some other countries in which he had negotiated.

The general atmosphere and the people created a much more open, comfortable, easy environment than anywhere in Eastern Europe. You didn't feel the pressure of "big brother" sitting on the wall and watching every move you made. Maybe they did have rooms bugged or phones tapped, but you didn't feel it the way you do in Eastern Europe.

Throughout the negotiations, frequent memoranda of understanding were signed. Essentially they recorded ideas on a piece of paper, which was subsequently signed by both parties. These memoranda were not agreements, but only mutual statements that discussions were still in progress, somewhat analogous to the minutes of the meeting.

Product and Technology Choices

A critical factor in the choice of initial product and technology to be transferred was that it be a good "learning" product. Foxboro's executives wanted to ensure that the selected product was within the capacity of the Chinese to assimilate, that it offered a learning opportunity for SFCL to develop the capabilities that Foxboro would want to use as the venture progressed into more advanced products, and that it could be transferred without transferring massive amounts of personnel support. As one executive said, "We wanted to guarantee the Chinese a success experience." And another executive echoed, "You have to learn to walk before you can start to run."

The initial product that was selected was the electronic analog product line, Spec 200, that had been in use for nine years. It was still being enhanced in development and, therefore, the Chinese could not brand it "old hat." Moreover, it was the kind of product that Chinese importers were buying at that time from external contractors for a number of plants. In addition, as Foxboro's executives repeatedly emphasized, it was a good learning product.

The Chinese were very well informed about Foxboro's product line and wanted the analog product that was initially chosen. However, very shortly after start-up of this product line, the Chinese began to push for digital technology, which they were very keen on acquiring. Foxboro, however, maintained that perspective that "analog" had to precede "digital," and that only after one technology had been absorbed would Foxboro transfer the more advanced digital technology.

Criteria used by Foxboro to choose technologies for transfer included market and economic feasibility; that is, given the potential market and the economics of the technology, did transferring the technology make business sense? Second, did SFCL have the ability to assimilate the technology? This was a function of the availability of skilled technicians and an infrastructure that could adequately assemble, test, and perform quality control and also provide customer application engineering and service for the new technology. For the effective use of process-control technology, engineering know-how of nonhardware applications was considered extremely important by Foxboro and was, in fact, described at times as even more important than the hardware component.

Capital Structure

SFCL was set up as a limited liability company with the liability of both partners limited to the amount of equity capital contributed by each party. The total capitalization was $10 million, with SIC holding 51 percent and Foxboro 49 percent. According to the contract, the proportion would remain unchanged, but the total amount of capital could be increased or decreased after five years of operation, according to the needs of the partners.

SIC contributed its share of the capital in Chinese currency (renminbi), cost of use of land; and fixed assets including buildings, facilities, and equipment. Foxboro contributed cash in U.S. dollars. Besides cash, other types of capital invested in SFCL were evaluated by an appraisal group selected by the partners. Neither partner was permitted to transfer its capital to a third party without the approval of the other partner, who had right of first refusal.

Taxes were to be levied on SFCL's profits according to PRC law. There was, however, a strong incentive for reinvestment of profits. For the first two years, reinvestment resulted in complete tax exemption; for the next three years, reinvestment resulted in a tax rate 50 percent of the normal tax rate.

Royalty Payments and Foreign Exchange

Foxboro did not use its technology as part of the capitalization for SFCL. Instead, it capitalized SFCL separately and negotiated a technology transfer agreement that required royalty payments by SFCL to Foxboro. Foxboro accepted the long-term objective of balancing foreign exchange for the joint venture, which was a challenge because of the large amount of imported components used in SFCL's products. Foxboro agreed, as part of the negotiated agreement, to work with other Chinese companies to develop local sources of components. In addition, Foxboro paid commission to SFCL for the sale of Foxboro-manufactured products in China. Foxboro also accepted responsibility for all sales of SFCL products outside China and agreed to purchase components and parts from SFCL.

Conclusion of Agreements

In December 1980, Foxboro signed two general agreements, one for a plant in Shanghai and the second for a plant in Guangdong.[6] At this time, Foxboro assigned a senior manager, Thomas Stuhlfire, the responsibility of further developing and implementing the Chinese agreements. Stuhlfire had been with Foxboro for 21 years and had worked in customer service, engineering, manufacturing, and product development. He had also been general manager of Foxboro's European operations in the 1970s. He described his charter as follows:

> What we had to do was pretty clear: Set up a factory to produce Foxboro products to Foxboro standards. The question was, "How do you get there?" We have a pattern for how we would do it in the United States, but I had to figure out just how much we would have to deviate to accomplish the same thing in China.

One of the first actions Stuhlfire took in this assignment was to organize a joint venture team in Foxboro, Massachusetts. The team included senior persons, usually the second in command, from each department. These individuals understood their part of the business and could commit its resources to the project. Also they were unlikely to be transferred within the anticipated two- to three-year start-up period. The total team included 17 people and covered all bases. They were kept advised as a team and were encouraged to establish their own communications with their counterparts on the Chinese side. (In the early days, Stuhlfire required that he receive a copy of every telex.) As a result, a

---

[6]The Guangdong agreement was not subsequently implemented; hence this case focuses only on the Shanghai joint venture.

sense of continuity was developed, and whenever a specific functional issue came up in the negotiations, a specific individual had functional responsibility for it.

The team had responsibility for developing a series of specific agreements covering different areas. The first item was an agreement called the "Contract," which was essentially an elaboration of the general agreement. It expanded the 10-page initial agreement to about 25 pages. Since the implementation regulations for China's 1979 Joint Venture Law had not yet been issued, the group fleshed out the agreement with every foreseeable detail that one might want to refer back to at a future stage. The details included the contract, the technology transfer agreement, articles of incorporation for the venture itself, purchasing agreements for parts and material, and training clauses. Developing the detailed agreements took about 15 months, despite the fact that the Chinese had assigned a member of the Foreign Investment Commission as a full-time member of the team to avoid constantly submitting items for approval to the Commission. To Foxboro, this action further reinforced the Chinese commitment to "getting the thing done, and done right."

The three issues that required last-minute debate and clarification were foreign exchange remittance, freedom for Foxboro to purchase material from other Chinese factories for foreign exchange balance, and the ability for SFCL to fire people for economic reasons.[7] In April 1982, all the agreements were signed except for the sales agreement, which at the time was not a pressing issue since production would not commence until early 1983. The group at that time was committed to making the plant operational in nine months.

Ed McIntyre, Foxboro general counsel, commented on the factors that had made the final negotiations effective:

> We sent the agreements that we had drafted at our end to them in China in advance. That way they could go through our drafts before they even came here and bring another draft that we could use as a starting base.
>
> Gleason had also asked them to come with as much authority to negotiate as possible and got a telex back from MMBI [Ministry of Machine Building Industry] that the head of the delegation would have complete authority to make decisons related to the joint venture. In fact, after the draft agreements were signed, they were approved by the Ministry with minimal changes in the prescribed 60 days.
>
> During the negotiations, I think the challenge was to overcome the distrust that existed on the part of the Chinese. Therefore, on issues that were critical to them, such as up-to-date technology, they would put a lot of redundancy into the agreement. They were also concerned about the prices the joint venture would be charged for the same reason. The issue seemed hard to resolve . . . until we came out with an idea. We said we'll show you the prices we charge our joint ventures in other countries and charge you the same. After that, there was no problem. They just wanted to make sure we were not taking advantage of them. . . .

---

[7]The labor relations section of the 1979 Joint Venture Law provided for firing only in cases of "employee misbehavior." Foxboro negotiated the possibility of layoffs in case of economic problems in the joint venture, such as in the case of a severe decline in revenues.

On our side, we looked for the spirit of the agreement rather than worry too much about the language. For example, there were phrases in the final agreement that were really not phrased very well in terms of English, but we just let it pass instead of spending a lot of time negotiating to correct the grammar.

A statement of the contract summary released by Foxboro after the agreement read, in part, as follows:

Under the contract, (SFCL) can manufacture specified Foxboro products—existing and to be developed—on the basis of Foxboro technology transferred to SFCL. SFCL will use the most advanced Foxboro technology to manufacture and sell in the PRC those selected industrial process control instruments and systems specified from time to time in a Manufacturing Plan approved by the SFCL's Board of Directors. . . .
SFCL also will perform systems and application engineering and sales, and will provide customer and maintenance services. SFCL's general objective will be to contribute to the development of the instrumentation industry of the PRC in a manner profitable for both investors.

Constitution of SFCL Board

As part of the detailed negotiations, the structure and responsibilities of the Board of Directors of the joint venture were defined. It was to be a nine-member board that would meet every April and October.[8] The first chairman was a Chinese, Yu Pinfang, the head of the Shanghai Telecommunication and Instrumentation Bureau.[9] Other Chinese members of the original Board included Gu Juchuan, president of the Shanghai Instrumentation Corporation, and Wu Qinwei, head of the Shanghai Institute of Process Instrumentation, a research institute.[10] There were also two Chinese representatives from the joint venture, namely Yang Tong, the first Chinese deputy general manager, and Wang Dawei, the chief accountant. All Chinese officials were selected by the Chinese side. The U.S. directors were Earle Pitt, Foxboro's chairman; Colin Baxter, president; Donald Sorterup, the first general manager of the joint venture; and Thomas Stuhlfire. The U.S. directors were appointed for three-year terms starting January 1983.

---

[8]In late 1986, the Board decided that one face-to-face meeting per year would suffice from 1987 on.

[9]The Shanghai municipal government's Telecommunications and Instrumentation Bureau, subsequently renamed Shanghai Instrumentation and Electronics Bureau, was a newly assembled bureau that had all the telecommunications, electronics, and so forth under it. SIC was subordinate to this branch of the Shanghai government. The fact that Yu Pinfang, from about three levels up in the hierarchy, was so closely associated with SFCL was taken to be a signal of the importance the Chinese placed on this venture.

[10]This research institute was affiliated with the Instrumentation Bureau of MMBI in Beijing, and was hence, considered a link to the central government.

The Board had the chief authority for managing the joint venture. The initial manufacturing plan, for example, was approved by the Board and covered a period of five years. Using the criteria of market demand and SFCL's production potential, however, the Board could modify it at any time.

Foxboro and the Chinese partner had jointly prepared a list of important issues that required the approval of two-thirds of the Board members, which meant that the approval of at least one Foxboro director was required. New product lines could only be added by a two-thirds majority vote of the Board. Increasing the personnel in the plant beyond the preapproved strength required majority Board approval. In addition, although an initial list of "important issues" for Board approval had been prepared, the agreement was structured so that any two directors could declare any issue "important," thus forcing a Board discussion and a two-thirds vote for approval. Thus, several checks and balances had been built into the Board's governance process for Foxboro to actively influence the evolution of SFCL's strategy.

The Board also had a key role in resolving conflicts.[11] According to the agreement, efforts would first be made to resolve conflicts amicably inside SFCL; the second level of appeal would be the Board. If the Board could not reach an agreement, the issue would be referred to the Arbitration Committee of the China Council for the Promotion of International Trade. If the issues were still unresolved, they would go to the Arbitration Council of the Stockholm Chamber of Commerce in Sweden for final arbitration, and the Chamber's judgment would be final and binding. But, as one executive remarked, "If an issue ever reaches the point of outside arbitration, the joint venture is dead anyway."

START-UP

Start-Up Personnel

In June 1982, Yang Tong, the main representative from Foxboro's investment partner and Tom Stuhlfire's counterpart on the Chinese side, (later appointed the deputy general manager of SFCL), came to Foxboro for two weeks with his interpreter and his financial advisor. The basic purpose of his trip was to agree on a proposed agenda of issues necessary to get SFCL started. Stuhlfire recalled how he introduced Yang Tong to Foxboro's operations:

> We had a lot of people from different functions come in and talk about what they did and why they did it the way they did. Some functions were very familiar, for example, manufacturing. Surprisingly, the personnel function did not hold a lot of mystery. . . . Different, but understandable. I was quite astonished. . . . But they just didn't have the foggiest notion about distribution. In China, the

---

[11]As of the end 1986, SFCL had never confronted an issue that could not be resolved by unanimous agreement.

factories produce something, ship it where they're told, and not even care about whether it ever gets there. . . . The purpose of the June visit, therefore, was to cover the range of things from the familiar to the totally unfamiliar and to agree on a basic agenda for the next Foxboro trip to China. . . .

In July 1982, Stuhlfire went to Shanghai with a ten-member delegation. The group consisted of eight Foxboro employees, including Don Sorterup and Ed Tarala, who were potentially to be the expatriate general manager and chief engineer, respectively, of SFCL, and their wives. This trip served as an exploratory trip for the Sorterup and Tarala families. In addition, the delegation worked on a joint venture implementation plan, and at the end of the three-week period, SFCL's Board of Directors met to approve the detailed plan of implementation.

The delegation spent time on capital valuation, essentially surveying the factory and equipment and selecting the equipment that SFCL would retain.[12] The equipment that was accepted was valued on a depreciated basis.

The group worked through several drafts of a labor agreement; the major hurdle was Foxboro's insistence on its right to fire personnel. As Stuhlfire mentioned,

> We got the right to fire in there at least to our satisfaction. But basically, as in all socialist countries, you have a tough, tough time even if you have the right to fire legally. So you have to be very careful about how many people you put on the payroll to start with.

The group also defined the organization structure in detail, including the number of departments with their titles and charters, the number of supervisors, foremen, and workers. Each of these issues was the subject of much debate because the Chinese side tended to overstaff dramatically. The plant that Foxboro took over, Shanghai Meter Factory No. 3, had six hundred employees when the Chinese ran it. The SFCL team finally agreed on an initial ceiling of three hundred people, beyond which operating management would require permission from the Board of Directors. Yang Tong, who had been general manager of the factory before the joint venture, selected the three hundred employees from the original six hundred. The rest were assigned to other SIC factories.

Foxboro's executives attempted to be more directly involved in selecting the senior staff. They reviewed the resumes of the Chinese senior staff and gave the Chinese side resumes of all the people Foxboro brought to SFCL. This was done to get the Chinese side used to not being secretive about qualifications and also to show that Foxboro was bringing in some of their best-qualified people

---

[12]The land associated with the joint venture had been valued in the earlier negotiating session and was already specified in the contract at this stage. Essentially, they had done a survey around Foxboro facilities worldwide, picked average numbers, and agreed upon a square meter value for which SFCL would pay rent.

to the joint venture. Even though Foxboro's executives could not evaluate the Chinese resumes effectively, they felt they had made their point.

The group also set up compensation policies for the three classes of people in SFCL: the workers, the senior Chinese staff, and the expatriates. Decisions about compensation for the workers and expatriates were relatively easy. For the workers, the Joint Venture Law required a wage between 20 and 50 percent more than in state enterprises in the same area. Thus the team took what SIC paid its workers in Shanghai and added the percentage premium. The expatriates were paid according to Foxboro's standard policy for expatriates. But the senior Chinese staff presented a dilemma. On the one hand, there was the agreed principle of equal pay for equal work; on the other hand, the multiple between the Chinese and U.S. professional salaries was in the neighborhood of 60. The team compromised somewhere in the middle. As one Foxboro representative recalled, "It was the price we had to pay for harmony with our partner. We gave them enough so they could report it with pride."

At the end of the three-week stay, the first Board meeting was held. This meeting approved the organization structure, manufacturing plan, remuneration agreement both for expatriates and Chinese staff, training program, initial capital injection agreement, insurance policies, selection of Chinese auditors, and other miscellaneous issues necessary to set up the company.

Stuhlfire recalled,

> We started with a clean slate and came out with an organized, staffed, registered company with a manufacturing plan, . . . partly because of a recognition early on that time spent in exploring assumptions, principles, and approaches before getting into nitty-gritty issues was time very well spent. So we spent time introducing people and functions. Here is the distribution function, here is how we organize it. We may or may not do it this way in China. . . .
>
> The purpose was to build a Foxboro factory in China, not a Chinese factory. So Foxboro had to lead the way, and they accepted that, but there is a "not invented here" and "we don't do things that way in China" syndrome, and we wanted to get over that in principle instead of confronting it when an issue required a decision. . . .
>
> Both Yang Tong and I understood that our job was to make something happen that required the melding of two cultures. I wasn't going to change China, and he wasn't going to change the United States. What we had to do was find ways of making something happen at the interface.

One of the first activities that Stuhlfire initiated at this time was the creation of an English-Chinese dictionary for the technical jargon that was an integral part of Foxboro. Being in a highly technical business, the instructors who would teach the Chinese engineers would talk jargon and the interpreters would not find the words in their dictionaries. So, during the first training visit to Foxboro, two SFCL employees, the training manager and the person in charge of documentation, used Foxboro's sources, industry sources, and Chinese sources to put together a joint venture dictionary. The English-Chinese glossary also became an attractive handout to customers. It was a very comprehensive glossary that the customers found helpful while reading technical literature in the field, and with the Foxboro logo on it, it was effective public relations.

The General Manager-Deputy General Manager Dynamic

According to the agreement, the first general manager (GM) of SFCL would be American and the first deputy general manager (DGM) would be Chinese. The Chinese side had, in fact, insisted that the first general manager be American using the logic that "if you're going to teach us, that's how it should be." After three years, however, the designations would be reversed with a Chinese GM and an American DGM. In spite of the difference in designations, however, the two executives cooperated and worked together as equals, with each individual's relative input depending on the particular issue on hand. Stuhlfire explained the GM-DGM roles:

> On the one hand, SFCL is a factory and has the job of building a product. But, second, it is a joint venture, half owned by each side. Third, it is a cultural experiment. And, fourth, it is a political symbol. What you do every day should keep all of these in mind. So, when it comes to running the factory, technical decisions about buying equipment, for example, the Foxboro people may have greater input. When it comes to financial decisions about how much discount should be given, you may see more of the Chinese DGM. When it comes to political and governmental issues, the Chinese head does his bit with the Chinese and the Foxboro head does his bit with the United States, and then there are many issues, such as personnel as well as many symbolic efforts, where we are hand in hand. . . .

One of the major challenges that Stuhlfire experienced in the beginning was to get SFCL recognized as a "real, legitimate Chinese company." As he explained, even though SFCL was a legal Chinese company, many Chinese still perceived it as "a bunch of foreigners in their midst." As a result, they were reluctant to respond to requests until the Chinese partner "put his stamp on it." Gleason added to this perspective:

> One of the reasons that I really don't understand companies wanting a 100 percent venture is I don't know how you'd get things done without a good Chinese partner. For example, when we got the first bid for redoing the plant, it was absolutely ridiculous. But we didn't have to fight through that ourselves. We sat down with the DGM and said, "This bid is ridiculous." The DGM said, "It sure is, let me have it." He went out and found a contractor who did the job at the right price. You really need your Chinese partner to get things done properly.

Don Sorterup, appointed GM-China in Foxboro in August 1982, and Ed Tarala, chief engineer, became Foxboro's first expatriate managers in China. Sorterup had joined Foxboro five years earlier when Foxboro acquired a small company that he managed. Sorterup was a generalist with a background in consulting. He had not traveled outside North America and believed, in retrospect, that his relative lack of knowledge about China worked to his advantage because he did not go to China with preconceived notions. As he commented,

> I didn't go there and try to be a Chinese. They want us for our way of working and managing and our technology. Therefore, one should go there and try to change only those issues that one has a charter to change and skirt around the others. Otherwise, you're setting yourself up for a lot of frustration.

Organization Structure

In the start-up phase, Stuhlfire and Yang Tong designed the philosophical basis of the organization, addressing questions such as why certain departments were necessary, how they would be expected to operate, and so forth. It was a functional organization, comprising personnel, financial, engineering, and production departments that evolved through several stages. In the August 1982 meeting, the Board of Directors had decided to let the GM-DGM hire enough people up front to meet the needs through the third year. The objective was to put the structure in place, staff all the functions, start training the people, and get them used to working together, rather than follow the growth path of a typical entrepreneurial organization. SFCL started with 286 people in 1983 and made a budget every year to set the work force ceiling for the year.

In addition to the formal structure, however, Foxboro's managers also had to instill a new sense of internal communication. The notions of internal communication—writing memos for the file and circulating papers "for your information"—were alien to the Chinese employees. As a Foxboro manager explained, "I had to take them out of a conference room one day and show them an in-basket and an out-basket to convince them we really used an internal mail system. But they are very good communicators, once the system is explained. You just have to get down to some very fundamental stuff and work your way up to ensure you've got a common ground."

There were many anecdotes describing how the performance of SFCL's employees changed after effective communication. Sorterup, for example, was dissatisfied with the clean-up job done on the tiles in the plant. As he described it, they would go around, wave a dirty mop, and say, "The job's finished." One morning, he put on his oldest blue jeans and a pair of old sneakers and went in with a bucket and scrub brush and got down on his hands and knees on the floor. He scrubbed a room and said, "Now that's what I want." From then on, they did an excellent cleaning job.

Initial Priorities

After being appointed GM-China, Sorterup stayed in Foxboro, Massachusetts, for the first three months coordinating training and other activities. He moved to China in January 1983. He described his initial concerns as follows:

> If we were going to transfer technology, the first thing was to create the facility to receive it. What that meant was not just documentation and mental understanding of our product, but to provide adequate systems and tools so we could measure them on their results. We also had to provide them a reasonable learning curve. . . .
>
> And I had to create a good facility. I strongly believe that everything should be neat and tidy. If you have a quality facility, you have a quality product.
>
> Getting all this across was a "teaching experience" and teaching by example. The contract documentation was excellent, but there were so many events and reactions every day that they could not be planned for. There was a lot of shooting from the hip. . . . As one person said, "The negotiation really begins *after* the contract is signed. It's the application that's tough."

When Sorterup arrived, he found SFCL's physical building, some old equipment, and 286 employees, but no work. Believing that it would be a bad way to start by having people sitting idle, Sorterup immediately initiated a "make-work" project. He put 35 people to work on building a "Shanghai-Foxboro Company Limited" sign to put on top of the plant. He took photographs of the progress every day and displayed them prominently. In retrospect, Sorterup commented that "the effort was a big morale booster and built a feeling that we're all in this together and we're going to do it."

As Sorterup worked with the Chinese engineers, he also found that they were very poor in problem identification. They tended to focus on symptoms and were weak in generating alternative solutions to a problem. As Sorterup explained, one of his key priorities became to communicate a philosophy and culture of problem solving:

> In the United States, we live a life of choice from the time we are born. We grow up on decision making. But Chinese children never touch the floor; they have everything done for them.
>
> So I had to communicate by example that it was OK to listen to others, to learn from peers, and to make mistakes. I used to run weekly middle management seminars. These were group meetings with a completely open agenda. I did it in a group to break their usual habit of doing everything one on one. And we would just pick up a problem and talk about it. Once we talked about a leaky roof and I asked them what the problem was. Seven out of eight people focused on symptoms, only one got to the problem, which was that it was a faulty design.
>
> I spent most of my time getting a feel for where everyone was coming from and looking for exceptions, because behind each exception would be a lot of issues. If someone was not wearing safety glasses, for example, it was an issue of individual safety but also one of clarity of communication, lack of proper education, inability of supervisors to take a stand, and so forth. So it was constantly a process of sniffing out these issues, bringing them to Yang Tong and saying, "I think we have a problem. What do you think we should do?" In the beginning, Yang Tong would not respond. I was the teacher so why was I asking him? It became

a question of getting dialogue going, of instilling the discipline of asking for something when you need it.

Technology Transfer

Foxboro started with an analog product line, the Spec 200; subsequently it added electronic transmitters, and in February 1985, received a technical data license for a digital computer systems–based product line, the Fox 300. Since materials and components amounted, on average, to 70 percent of Foxboro's product costs, SFCL had 30 percent value added through labor and overheads from day one.

The nuts and bolts of transferring technology had two components: paper (documentation) and people (training).

Documentation. The paper component of technology transfer, namely, technical documentation, involved a series of administrative, cultural, and governmental hurdles. The first step in transferring technology was to obtain a technical data license from the U.S. Department of Commerce. Once a license was received, Foxboro established a system for tracking the drawings and specifications that were being transferred against the license. Every time drawings were sent to SFCL, they went with a listing showing the details of the license against which they were being sent, and SFCL had to return a copy acknowledging receipt.

At the Shanghai end, SFCL organized a documentation control center that received and maintained all the documentation received from Foxboro. They maintained the drawings and did the official translations required, whether they were technical or sales instructions. The center also maintained the information already in the public domain for which no government clearance was necessary, such as price sheets and product sheets. Anyone at the Shanghai location could look up these records at the documentation center. By mid-1986, SFCL had translated 5,000 Foxboro technical drawings and 200 user documents into Chinese.

The situation at Foxboro's end became somewhat more complicated with each new license. When a new product was added, Foxboro could either amend an existing license or add a new license. For computer products, Foxboro obtained a new license. It, therefore, needed to differentiate the information being sent against that license as opposed to other licenses and to eliminate the common parts. Since the SFCL venture was for 20 years, Foxboro preferred to use amendments instead of constantly applying for new licenses.

In addition, since Foxboro had entered into a dynamic technology transfer agreement, frequent applications would need to be made for the transfer of technical data. License applications for technical data transfer could not be made until all the product development and design documentation was complete, whereas, through an entirely different process, licenses to ship to China complete products manufactured in the United States utilizing the same technology could be secured almost immediately. Obviously, after applying for the data transfer license, there was a further lag in the approval process before the design documentation could be actually transferred. Apparently, the U.S. government approval process for product export licenses had been streamlined, but the process for design and data transfer licenses was still fairly cumbersome.

Training. Training was the second major component of the technology transfer process. Since training was also subject to license control, the detailed content of the training had to be approved by the U.S. government. In addition, SFCL had to submit the content of the training program to the Chinese government and then have a list of the trainees approved by the Foreign Affairs Bureau before U.S. visas could be obtained.

First training program. In October 1982, 20 of the senior Chinese administrative and technical staff came for nine weeks of general management training and specific training in their specialties. In the first few weeks, the trainees were exposed to Foxboro's organization and management philosophy. Notions as basic as delegation, authority, and responsibility prevalent in Western organizations were discussed because Foxboro executives had observed the extreme centralization of power in Chinese factories. For example, sometimes they have only one telephone, which is kept in the GM's office; often *all* letters are sent out through the GM. Since much communication was expected to transpire between Foxboro and Shanghai, communication problems were likely to arise unless SFCL reasonably reflected the organization and culture of Foxboro and its other associates.

The general orientation was followed by on-the-job training in each trainee's functional specialty. Larry Martin, the accounting and finance representative on Foxboro's joint venture team, was in charge of the training in accounts. Of the two accounts trainees from China, one had already been designated as SFCL's chief accountant; the other as the second-in-charge. Martin took them through functions such as receivables, credit, and cost accounting with the ultimate objective of structuring an accounting system that would be a compromise between the level of their knowledge and the needs of the joint venture from Foxboro's perspective. Before leaving Foxboro, they prepared SFCL's operating budget for 1983, including cash flow in both renminbi and U.S. dollars, based on group sales projections.

Fred Morse had been part of the original Foxboro team that visited different plants in China, and he now coordinated the manufacturing and engineering component of the training. Foxboro had brought over the SFCL engineers responsible for process engineering, ordering the material, quality control, and quality assurance.[13] Most were graduate engineers in their late thirties, and 80 percent had prior experience in similar products. They were in Massachusetts

---

[13]Quality Control designed the inspection procedures required on the factory floor. Quality Assurance determined whether the manufactured products met their specifications.

for 12 weeks and went through every step of the product. The process engineering trainees, for example, first learned about Foxboro's design philosophy and then were asked to design and build a test product. They were required to build their own work sheets and develop their own procedures before returning to SFCL.

Foxboro had set aside some production areas that simulated Chinese conditions. For example, these areas did not have automatic forming equipment and automatic insertion equipment because these would not be available in China. Here SFCL's trainees did hands-on assembly and testing and were even given defective products for trouble-shooting.

Foxboro's managers repeatedly emphasized one point: "We want you to get sufficient training so that you become the trainer when you return." In Chinese factories, individuals tend to be secretive about their knowledge. Foxboro did not want these trainees to "return to China and lock up their books." They hoped to encourage sharing by saying that everyone could not have the same training so that they would have to share and train one another.

Foxboro put much effort into making the first training experience effective. All instructors were briefed on the nature of the joint venture with the Chinese and on Foxboro's obligations. Instructors were also given a monograph on working through interpreters, since all the training was with simultaneous interpretation.

Initially, the Chinese students were taught the basic concepts of training in the United States. It was emphasized that asking questions was not considered an insult. In addition, a very structured evaluation process was designed to monitor the training on an ongoing basis. Every class had a student leader who, at the end of every session of every class, would confer with his or her colleagues and fill out a simple, one-page bilingual form on how well the session had gone, what questions they still had, and so forth. The instructor would fill out a similar form. The forms were collated by the training manager at the end of each week and reviewed at a management meeting with the Chinese and U.S. representatives at the start of the following week. Over time, as the training went well, the evaluation became less formal, but weekly sessions with the delegation leader were still continued so midcourse corrections could be made whenever necessary. The only consistent feedback from the Chinese trainees was that they wanted to go deeper than Foxboro wanted to go at that time. The Foxboro company line was, "Get this accomplished and we'll go to the next step."

The Chinese trainees returned to Shanghai in December 1982. Two weeks later, Foxboro's two expatriate families went over.

Subsequent training. The second training session, given in summer 1983, lasted three months. Six sales and service trainees (four in sales and two in service) came to Foxboro from the associated Sales and Service Company (Shanghai Foxboro Sales & Service Co.), a separate company wholly owned by the Chinese partner, SIC. This team comprised the sales manager, the deputy sales manager, and four other key people. The first week of the training focused almost entirely on the concept of customer satisfaction with the remainder on market analysis and selling strategy. Immediately after returning to China, they started to call on potential customers.

In September 1983, the next group of SFCL trainees arrived. They consisted of 15 individuals from one level down in the organization. This was highly specific training, all within their own specialties, and included finance and accounting, quality control, engineering, and manufacturing.

By this time, the China project was well into its start-up phase, and Stuhlfire gradually withdrew from active participation. However, because of his subsequent position as head of Foxboro's training facility, which trained customers and internal people, Stuhlfire continued to be involved with the Chinese trainees. During his two years of primary involvement, Stuhlfire made six trips to China and spent an average of three weeks each time.

In commenting on Foxboro's overall approach toward training their Chinese partners, Stuhlfire said,

> The two issues are the effectiveness of a given means of technology transfer and expense. What we're basically doing is the front-end training here and then moving the training to China. Front-end training is done here so they can get out of their Chinese element, get immersed in Foxboro, and see who we really are and why we do what we do. They're typically here about three months, so they get to know us well, and we get to know them. Now many people here know people in Shanghai. This serves as a base so we can now push training in Shanghai. It's a lot cheaper to send one instructor there than to bring 30 students here, particularly considering the foreign exchange issue.

In 1984, a group of engineers came to study computer products and their means of assembly. The next year a group of product engineers came to study existing products in greater depth and look at some new products. Foxboro averaged about one major training program per year for SFCL trainees in Massachusetts. In addition, on-site training in China was expected to increase. Fred Morse described a recent on-site training experience:

> We sent a systems engineer there, and it was the best experience yet for the company. They had an existing project and he worked with them directly on it. When he saw a problem, he would get them all together and tell them, "Look, this is how you approach such a problem." At times he would lecture to 40 people at a time. There was a great multiplier effect. It's more cost effective to go there. But you almost need to be a generalist to go there because they ask you all sorts of questions. This is a problem for us in sending people over there, because we are essentially a company of specialists. We may have someone who specializes in test procedures for a certain type of board, but they have ten types of boards and it's too expensive to send ten people there.

Foxboro also set up a training department in China so SFCL had a unit responsible for training, especially in areas such as workmanship training, soldering standards, mechanical assembly, and so forth. SFCL's training supervisor spent time in Foxboro, Massachusetts, putting together training courses using videotapes, slides, and so forth. The courses were then conducted in China.

In January 1984, Foxboro wanted to make the SFCL plant capable of manufacturing under Factory Mutual Approval for Intrinsic Safety, which is a classification system based on Factory Mutual inspection of plants and procedures to ensure product safety. Factory Mutual examined and passed the SFCL plant, which meant that a lot of procedures, especially quality control, documentation control, and so forth, were properly implemented.

By 1986, over 125,000 man-hours of training had been carried out, including more than 45,000 man-hours spent training 69 people outside China.

Transferring technology to organizations other than SFCL. Because of the unique characteristics of China's infrastructure, Foxboro also transferred technology to organizations other than SFCL. Due to the lack of locally made components meeting Foxboro's quality standards, for example, Foxboro's engineers would visit potential supplier organizations, review their systems, help implement changes in their methods, and even help them contact possible foreign collaborators. Foxboro also transferred technology to the research institutes that were affiliated with SFCL's potential customer industries. For example, a Power Research Institute would do the overall project design for a power plant in its jurisdiction. Foxboro would therefore help the Power Research Institute develop expertise in control instruments that the Institute, in turn, could factor into its plant design.

Marketing Strategy and Organization

Sales and service organization and personnel. SFCL did not perform sales and service. Instead, a separate Sales and Service Company (SFSSC), wholly owned by the Chinese partner, SIC, was the exclusive international representative (IR) in China for all products manufactured by the joint venture and also for all other products sold in China by Foxboro. This arrangement had been specified in the original joint venture agreement. Foxboro had IR arrangements in 51 countries where the market potential did not justify a more direct investment and presence. In China, the situation was essentially similar except for the fact that the IR represented both the joint venture and Foxboro. Yet it was closely "married" to the joint venture, since it was "owned" by Foxboro's Chinese partner. It also reported to Yang Tong, SFCL's first DGM and later its GM.

Ernest Debellis, a Foxboro manager with extensive experience in marketing and applications engineering, had become involved with Foxboro's China operations in the summer of 1982, after the agreement was signed, to help with product planning and market-related issues. During his first trip to China in August 1982, Debellis visited customers and became familiar with the technological sophistication of the users and manufacturers. As he described it, for a high technology product such as Foxboro's to generate customer satisfaction, an engineer had to work closely with customers and train them to use the product properly, but "there were many Chinese engineers who had never even been inside a customer's plant." Concepts of warranty and after-sales service did not seem to exist. The initial marketing training at Foxboro was designed to overcome this lack of customer orientation.

After coordinating the initial sales training, Debellis visited China about every three months as "coach" to SFSSC to reinforce the customer orientation and provide on-site help in implementing an effective marketing strategy. Although Debellis did not have an official position with SFSSC, it would call Foxboro if it had any questions or problems. SFSSC also regularly sent Foxboro a monthly major projects report for potential projects over $100,000 and the project closure rate.

Subsequent training outside China for SFSSC engineers included six people in Singapore in the summer of 1984 and another four people in the United States in the summer of 1985. Although training in Singapore was much less expensive for the Chinese, some of the training had to be done in the United States, where certain product capability only existed.

The Sales and Service Company, which started in 1983 with 10 employees, had grown to 45 employees by 1985. It was located in a new building in the same compound as SFCL and had an extensive demonstration center for Foxboro products as well as a training center for customers. Training courses were run every month for 25 to 30 customers at a time on the product manufactured in China. Products that developed problems in the field were first handled by SFSSC and were passed on to SFCL if SFSSC lacked the necessary equipment or skills.

Marketing and competition. Foxboro's basic marketing thrust was to emphasize its strengths in petrochemical, fertilizer, and electrical power-generation industries. They had analyzed the major markets and targeted key accounts. One of the employees in SFSSC, who worked as an advisor, had access to key people in most of the major plants and helped set up meetings for the salespeople.

The most distinctive and different feature of the Chinese market was that the economy was planned rather than market driven. This feature led to interesting observations. For example, because many instrumentation plants had production quotas with little incentive to exceed the quota, SFCL would experience a sudden increase in demand as the quotas for the period were achieved by their Chinese competitors. (This aspect, however, was expected to change as profit goals became more of a norm among Chinese companies.) In addition, much of the customer decision making was centralized. SINOPEC, for example, had centralized control over 39 oil refineries, which made them analogous to an Exxon in scope of projects. Such factors made the Chinese market in some ways easier to deal with than the U.S. market. SINOPEC, for example, was willing to give out information on future projects because they were not worried about competition, whereas an Exxon would be very secretive because of competitive reasons. Also, people in the ministries had a lot of information they were willing to share.

SFCL-manufactured products essentially competed with several Chinese factories, some of which were actually owned by SIC and had products that were functional equivalents to those of SFCL. When the Chinese customer had foreign exchange, however, SFCL competed with every company in the world. The Chinese thought about buying in renminbi and dollars in completely different ways, and Debellis estimated that about 75 percent of the total market in China for control instrumentation products would be sourced from within China in renminbi. Over the long term, therefore, SFCL products were expected to become dominant relative to Foxboro, USA, products. Foxboro's executives believed that the way to win in China was to sell a product in renminbi that was technologically competitive with international technology.

MANAGEMENT OF THE ONGOING PROCESS

Plant Labor Relations and Party Presence

Each plant in China had a labor personnel department that independently measured the performance of the employees. This measurement was quite independent of the measurement that SFCL did for its internal incentive program. This department also ran the party meetings that were held every Tuesday morning. All employees would come to work an hour early and go into the cafeteria where a speaker from the party would address them.

Human Resource Systems and Policies

Some joint ventures in China paid a government agency a lump sum of money equivalent to the cumulative payroll, and the agency, in turn, distributed the wages, deducting an amount for the government, which took care of housing and other benefits. SFCL, however, paid its own payroll, similar to State Enterprises.

Among the programs that SFCL instituted was tuition reimbursement for job-related education. There was one case, however, when an SFCL employee was told by the Labor Relations Department that he would not receive compensation for missed time, even though Don Sorterup, the GM, had approved his education-related absence. Sorterup established his authority in this case after the employee informed him about it. The fact that the Chinese employee had informed Sorterup of the problem was unusual, since the Chinese were accustomed to seeing the labor relations people have the final say. This issue also reflected the amount of trust that SFCL's top management had been able to develop in the company.

SFCL was also the first joint venture in China to institute an individual performance program to award bonuses to employees for good performance. This program evaluated employees twice a year on such criteria as output, attendance, good work habits, and quality. This program, however, was not publicized extensively outside the joint venture because of Chinese concerns about how it would affect the work ethic in state-owned enterprises.

Foxboro had also managed to change recruiting habits. As Sorterup recalled, one Friday shortly after he had arrived in China, he received a phone call from SFCL's Human Resources manager informing him that SFCL had been allocated four engineers by the Shanghai Instrumentations and Electronic Bureau (SIEB) and they would report to work on Monday. Sorterup's response was that, first of all, SFCL did not need them, and second, SFCL ought to interview any new recruits before hiring them. The Human Resource manager's first reaction was that SFCL ought to take the people anyway so that they would be available when and if SFCL did need them. After Sorterup put his foot down, however, the engineers were not hired. The next July, SFCL was given advance notice that people would be available. Yang Tong and Sorterup interviewed them and selected accordingly. Such an approach to recruiting was very different from other Chinese companies, which were usually allotted people and had little or no choice about who was hired.

SFCL also replaced people whose performance was consistently unsatisfactory. This was, however, a long process because the individual first had to be found a new position. In addition, SFCL also wanted all employees, unless they were obvious misfits, to be given a chance to prove themselves. In general, problems with performance were rare at SFCL. The executives believed that SFCL had received very good people because the Chinese partners wanted the joint venture to succeed as much as Foxboro did.

Accounting and Control Systems

For general accounting, SFCL maintained the books of the company using double-entry bookkeeping in a manner similar to that of Foxboro. The process was, in fact, simplified because SFCL, like other Chinese entities, paid for goods before they were even shipped. Therefore, accounts receivable and accounts payable were not a consideration.

Cost accounting and fixed-asset accounting, however, required some reorientation on the part of SFCL's accountants. The Joint Venture Law was relatively new and hence permitted a lot of flexibility in the choice of depreciation methods, but at the very outset SFCL's accountants structured the system so that it was beneficial to the Chinese government rather than to the enterprise. As Martin remarked,

> They hadn't come to the realization that private enterprise meant using the law to the best advantage of the corporation. In their society, they had always used

the law to aid the government. . . . This mentality was the real challenge during the training.

In situations where the law was general enough, Foxboro's managers would choose the option most advantageous to SFCL. For example, the accelerated depreciation method would be chosen when possible because it was more advantageous to the joint venture and because the law didn't prohibit it. Since SFCL's accountants had only used the straight-line depreciation method, Foxboro managers would reassure them about using the accelerated method by having them confirm it with the appropriate Chinese government agency. Over time, SFCL's accountants became more open to new ideas. This trend was expected to continue because Chinese party leaders were placing greater public emphasis on "profits."

Foxboro also transferred the "technology" of inventory control methods. Initially, SFCL used manual stock cards for inventory; there was much duplication in record keeping because the accounting people, the materials people, and the production people were all maintaining stock cards. From a technology transfer point of view, inventory control was experienced as quite problematic. Foxboro in Massachusetts had been on a sophisticated computerized MRP system for many years and found it hard to train people on a manual system. Ultimately, SFCL acquired a fairly powerful IBM computer.

Martin traveled to China in October 1983, about nine months after the accounting trainees had returned to SFCL, to follow up on the training. During this period, SFCL started to manufacture and produce monthly budget reports for management, copies of which were also sent to Martin for comments on system modifications. Martin found that SFCL's general accounting functions, especially the cash flow and foreign exchange projections, which was its life blood, to be in excellent shape. But its systems for work-in-process inventory and product costing were weak. Martin worked with the SFCL accountants in developing control systems in these areas. He concluded that SFCL had reached an operating level where "more would be gained by sending someone from Foxboro to China to work hands-on than to have them come to the United States to listen to concepts."

Information Reports

For Foxboro's record keeping, SFCL was a joint venture with a minority equity investment. Since it was not a subsidiary, Foxboro did not consolidate SFCL's results and hence did not require a profit and loss (P&L) or balance sheet. A telex about the total income would be sufficient for Foxboro to factor its 49 percent into the corporate income statement. However, to obtain more visibility at Foxboro corporate, Foxboro asked for the P&L and balance sheet to be sent regularly. Foxboro also required SFCL to send a set of monthly reports that it usually reserved only for its subsidiaries. These informational requirements included a detailed breakdown of all operating and other expenses for the month and year to date. Business plans and targets were basically developed at SFCL's Board of Directors.

Capital Investments and Foreign Exchange

Investments and payments by Foxboro took several different forms. Foxboro received royalty payments for its technology based on sales. Foxboro paid its share of the rent for the land in U.S. dollars, as part of its capital injection. In addition, Foxboro also continued to inject more dollars over the course of the venture because most of what SFCL had to buy had to come from the United States. As a result, the capital injection from both sides continued to climb. As one Foxboro executive mentioned, "We have loads of renminbi but no dollars."

SFCL started to address this imbalance by negotiating contracts with Chinese customers requiring payment in U.S. dollars whenever these customers had foreign currency reserves. In addition, SFCL searched for labor-intensive products to manufacture in China and export to Foxboro, USA. The first such product was instrument transformers, which SFCL manufactured, with some imported components, for supply to Foxboro in July 1985. These strategies, combined with efforts to increase manufacture in China, were expected to reduce the foreign exchange problem over time.

SFCL maintained two bank accounts—a U.S. dollar foreign exchange bank account for imported materials purchases and a renminbi account for labor and some materials purchases. The U.S. expatriates were paid partly in renminbi, for local expenses, and partly in U.S. dollars. Foxboro made the dollar payments and charged it back to SFCL.

Business Visibility

On a few occasions, Foxboro won contracts in China even though it was not the lowest bidder. One executive hypothesized that this was because Foxboro had acquired so much visibility through its participation in SFCL. In addition, Foxboro's ability to service its Chinese customers out of Shanghai made it easier to rationalize a premium price.

Some of the products that Foxboro sold through these contracts were manufactured only in the United States. These products were exported through SFSSC for payment in dollars.

Foreign Exchange and Local Sourcing

The joint venture contract explicitly stated that "when possible, raw material, parts, and subassemblies will first be supplied by Chinese vendors; that portion of such material which cannot be obtained in China shall be imported from Foxboro or some other outside source." Here China's industry infrastructure posed a problem in developing a local supply network. Chinese bureaus were vertically integrated; for example it was hard to find a castings manufacturer that made castings for a lot of different customers. Manufacturers' quality standards were typically determined by the bureau governing their activities. They were evaluated on meeting the state plan and not on criteria such as delivery, price, and service, which SFCL imposed.

In addition, Foxboro faced a "capacity problem." SFCL in particular, and the instrumentation industry in general, was not a large user of materials from the standpoint of economies of scale in manufacturing semiconductors, special alloys, and so forth. Investing resources in building such supply capability, therefore, could not be justified on a cost-benefit basis because the market size was not big enough to support efficient production scale.

Foxboro found it difficult to find local components that met the military standards it used. For example, Rathke, SFCL's chief engineer in 1986, had to write to 20 steel mills to inquire if they would work on producing the specifications that Foxboro needed. For plastics, Rathke went to a mold factory to get the molds made; he imported the plastic raw material and then took the molds and the raw material to another company to make the plastic components.

As SFCL's volume increased, so did its foreign exchange deficit since the majority of what they were making in China, even in 1985, was assembled from imported material. As the project progressed, the Chinese government accommodated SFCL and did not demand that the foreign exchange deficit be "fixed." In spite of the government's patience, however, Foxboro felt a high degree of self-imposed pressure to balance the joint venture's foreign exchange and continued to work with potential local suppliers on qualifying their components.

By 1986, Foxboro had qualified several components including stainless steel wire, resistors, sheet steel, aluminum, and some laminates and circuit boards. Foxboro was very strict in insisting on its standards because all its designs, worldwide, used the same standards and specifications and Foxboro had agreed to produce to its worldwide standards at SFCL.

Relationship with Partner

Central to managing the joint venture for long-term success was effectively relating with the senior Chinese managers at SFCL. Sorterup described his experience with Yang Tong:

> Yang Tong had a much harder job than mine because he had multiple bosses. When it seemed that he was arguing against me, I had to realize that what he was really looking for was help in satisfying his different bosses who could have conflicting drives and wishes. And he was really arguing every one of their cases. . . .
>
> I realized that when I gave an instruction to someone, he had to convince other people to get it done. If it was in the same work unit, he had a chance. But if he had to get something from a guy who was part of a different infrastructure, he had a problem. So I learned to constantly ask myself, "Am I asking for results that are achievable?"

As Foxboro's executives stressed, the problems they faced in China were never problems with their partner; the real problems were *bigger* than both of them and had to do with the general infrastructure, such as foreign exchange controls, local sourcing, bureaucratic inflexibility, and so forth. Rathke, who took over as chief engineer from Ed Tarala, also emphasized that "what you can do in the United States in two hours on the phone takes two to three weeks in China." Contributing to the frustration were the inefficient telecommunication and transportation systems combined with an orientation that required one to "meet in person to do business *after* setting up a meeting by letter."

Systems Phase

Foxboro made a distinction between its instruments and systems. An instrument was a stand-alone product, produced on a demand forecast; a system was usually a complex combination of instruments and computers, hardware and software, put together in a unique configuration to meet a customer's specific requirements. A system was usually big, complex, and not fully defined at the time the customer's order was placed. Consequently, a systems order required a continuing dialogue between Foxboro and the customer, totally different from receiving a parts order and shipping them from off a shelf.

Foxboro waited to transfer systems know-how to SFCL until the latter demonstrated the ability to produce basic instruments. However, by mid-1984, Foxboro agreed to go ahead. A group of ten systems engineers came from SFCL for training in mid-1984 for a combination of school training and practical training in Foxboro's systems division. At this time, SFCL also hired 17 graduates from Shanghai Technical University. One Foxboro manager described them as the best of the entire graduating computer science class.

In June 1985, SFCL shipped its first two sets of Fox 300 Digital Computer Systems to a chemical plant and an oil refinery.

FUTURE PLANS

Management Transition

In October 1985, Donald Sorterup, the first general manager, and Ed Tarala, the first chief engineer, returned to Foxboro. The new head of Foxboro in China was Ernest Debellis, who had worked with Foxboro for 23 years, mostly in applications engineering and marketing. He had visited China with the August 1982 Foxboro delegation and had already made several trips to help the Sales and Service Company. His background represented an important shift in the life cycle of the SFCL joint venture from a production to a marketing orientation.

Debellis described how he saw Foxboro's competitive position in China evolve during his tenure:

Our competitive advantage is to put the latest technology into the manufacturing plant, so we offer a product with up-to-date technology in Chinese currency. The half-life of technology in our field is three to four years, so we have to keep transferring technology to maintain an edge. That would take care of market and competitive issues, although it does leave the problem of balancing foreign exchange. That is what I see as my primary goal, namely, to get the right product manufactured, as much as possible, in China. SFCL has very limited ability and resources to expand its product line, so the products we choose is critical in terms of long-term positioning and market share. Our recent decision to go with a critical system product was a very important strategic decision. . . . In effect, we want to sell products that are integral parts of systems rather than just peripheral components.

Expectations and Long-Term Plans

The financial results of the joint venture had been quite promising. In its first full year of operation, it had made a small profit. It exceeded its 1985 sales forecast of 15 million renminbi, about $5 million. However, because almost all the earnings were in renminbi, the joint venture was constantly in need of U.S. dollars for the imports of materials and components.

In any case, Foxboro expected limited returns, beyond the royalty payments, from the joint venture in the first five years of its life because there were tremendous tax advantages to reinvesting profits for the first five years back in the business. Foxboro viewed its reinvestment in SFCL in the context of its 20-year joint venture agreement: It was getting in on the ground floor of the development of a major market and contributing to that development by building materials and products to enhance China's economic growth. SFCL would be "Foxboro of China," and Foxboro's management expressed the hope that its 20-year agreement would be renewed.

Postscript

A report from the New China News Agency (1987a) stated that the Shanghai-Foxboro Company had been enjoying a steady growth of profit and foreign exchange funds in its four years of operation. Its "business value" (revenues) reached about $7.5 million in 1986, a 37 percent increase over 1985, and its profits almost doubled. Its income and expenditure in hard currency balanced in 1986 because the Chinese government allowed its products to be sold for foreign exchange in China. At the same time, the company used more raw materials sourced in China and expanded exports. In the past four years, the company had sold 140 sets of automatic control apparatus and meters to power-generating stations and petrochemical and metallurgical plants in China. In 1987, the company planned to put three new products into production.

EXHIBIT 1   **Shanghai-Foxboro Company Limited (SFCL): Key Events**

Date	Event
1978	Senior management delegation from Foxboro visits PRC
1979-1981	Three exploratory trips by Foxboro, Massachusetts, experts to PRC and two investigatory trips by Chinese experts to United States
1980	17 Chinese engineers trained for six months at Foxboro for technical support of Foxboro products imported into China
December 5	General Joint Venture Agreement signed
1982	
April	Joint Venture Contract signed
June	Yang Tong, SIC representative, and first SFCL DGM visits Foxboro with financial advisor for two weeks
July	10-member Foxboro delegation covering all functions plus first Foxboro expatriates visit China for three weeks
August	First Board of Directors meeting and approval of senior management appointment and operating plans
October	20 SFCL management and technical staff for nine weeks training at Foxboro
1983	
January–March	286 people employed, including 106 graduate engineers and technicians; expatriates Sorterup and Tarala arrive from Foxboro
April	Spec 200 System assembled by SFCL and displayed at Shanghai Exhibit; second Board of Directors meeting; plant opening ceremony with Shanghai Vice-Major Xin Yuanxi presiding
June	First major Spec 200 system shipped to Liao Yang Petrochemical Fibre Corporation; six Sales, Service, and Systems engineers for three months training to Foxboro
September	15 Technical, Distribution, Quality Control personnel to Foxboro USA and Canada for three months training
October	Third Board of Directors meeting and approval to proceed with technology transfer of digital products; international sales agreement signed between Foxboro, SFCL, and SFSSC
1984	
January	Factory Mutual Corporation certifies SFCL in accordance with USA standards making it the only company certified in China
March	Fourth Board of Directors meeting
April	President Ronald Reagan visits SFCL
June	Twelve Systems and Product engineers for 12 weeks training in digital computer systems products to Foxboro
July	Six Sales and Service personnel to Foxboro, Singapore, for 80 days training
October	Fifth Board of Directors meeting at Foxboro; Systems Engineering department consisting of 25 engineers and technicians established
November	Three production lines—sheet metal fabrication, transformers, and transmitters—put into production

EXHIBIT 1 (*continued*)

Date	Event
1985	
February	Assembly of 800 Series transmitters started; technical data license for digital systems line received
May	First annual technical seminar held by SFCL, SFSSC, Foxboro, and SINOPEC
June	First two sets of Digital Computer Systems (Fox 300) assembled and shipped to a chemical plant and a refinery; 12 Product and System engineers trained at Foxboro for three months in digital product
July	First export to Foxboro of instrument transformers manufactured by SFCL: four Sales and Service personnel trained at Foxboro for service of analytical instruments
August	Term of office of first expatriates expires; new GM is Chinese with Foxboro expatriate as DGM
September	Three Systems engineers for Power Plant Control System and two Service engineers for Fox1/A Computer System for three months training to Foxboro
October	Presidents and chairmen from 36 of largest U.S. organizations visit SFCL to observe successful high technology joint venture in operation
December	Large Spectrum (Distributed Digital) system completed for petroleum refinery project
1986	
January	Three Service engineers for three months training in digital systems to Foxboro
March	Two Systems engineers and two Service Engineers to Foxboro Singapore for three months training in digital products
April	Premier Zhao Ziyang accompanied by Rui Xinwen, Secretary of Communist Party of Shanghai, visit SFCL in recognition of its achievements

EXHIBIT 2 Foxboro Company: Financial Highlights

	Thousands U.S. $	
	1983	1984
Net sales (shipments)	$553,102	$515,856
Income before taxes	6,377	972
Taxes on income	(1,600)	(3,300)
Net income	7,977	4,272
Earnings per share	0.65	0.35
Orders received	485,000	553,000
Backlog	216,000	253,000
Research, Development & Engineering expenses	45,894	49,402
Capital expenditures	19,600	25,504
Total assets	432,977	470,439
Shareowners' equity	310,591	291,985
Book value per share	25.14	23.64
International operations:		
Europe		
Sales	115,639	113,642
Income before taxes	7,079	1,010
Assets	85,367	81,672
Others		
Sales	72,191	64,390
Income before taxes	8,496	5,430
Assets	51,639	53,810

Exhibit 3 The Foxboro Company simplified organization

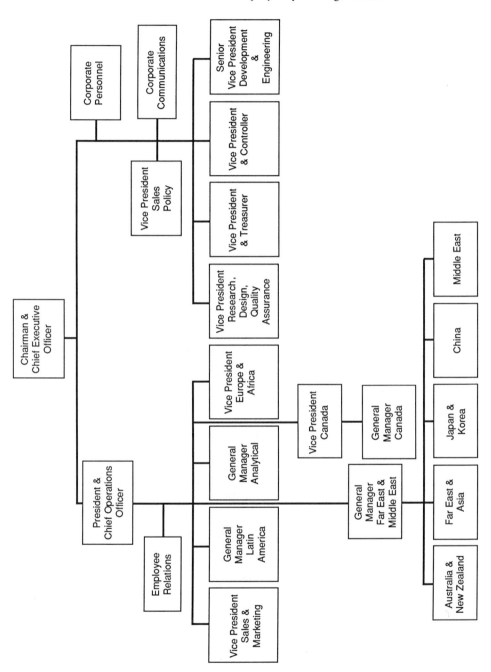

CASE 2

Shanghai-Foxboro Company Limited: The Chinese Perspective

INTRODUCTION

This case is based on an interview trip to China in June 1986, during which the research team visited the Ministry of Machine Building Industry (MMBI) in Beijing; and the Shanghai Instrumentation Corporation (SIC), the Shanghai Instrumentation and Electronics Import and Export Corporation (SIEIEC), and Shanghai-Foxboro Company Limited (SFCL) in Shanghai. The purpose of these visits was to investigate the Chinese view of the joint venture agreement concluded in April 1982 between the Foxboro Company of Foxboro, Massachusetts, and the Shanghai Instrumentation Industry Company (SIIC).[1] Discussions were held at MMBI with representatives of the Instrumentation Industry Bureau and other bureaus.

DEVELOPMENT

The need for more advanced process-control instrument technology arose in the context of the ambitious growth plans of China's basic process industries, such as electric power, oil and gas, and chemicals, for the years 1980 to 2000.

---

[1]The Shanghai Instrumentation Industry Company (SIIC) was subsequently renamed Shanghai Instrument Company (SIC). By 1987, SIC, which was a conglomeration of several production facilities primarily in the instrumentation industry, was dissolved as part of the government's drive toward decentralization. More autonomy was given to each production facility.

Process-control instruments are critical for maintaining output and quality in these industries. However, in the early 1980s, the Chinese instrumentation industry was still building control instrumentation based on the technology of the 1940s. Considering that the half-life of process-control technology in the United States in the 1980s was three to four years, Chinese officials and engineers believed that the Chinese instrumentation industry had a lot of catching up to do.

Shanghai Instrumentation Corporation

The Shanghai Instrumentation Corporation (SIC) was administratively subordinate to the Shanghai Instrumentation and Electronics Bureau (SIEB). It was a conglomeration of 53 subsidiaries that manufactured process-control and electrical instruments in Shanghai. The subsidiaries, in aggregate, had more than 35,000 employees. The average proportion of technical employees to total employees in SIC factories was 11 percent, ranging from a low of 8 percent in one factory to 38 percent in SFCL. According to Zhu Ruifen, DGM of SIC, the director of each factory was appointed by SIC but each factory was independent, with its own legal and finance personnel.

Out of the 53 subsidiaries under SIC in mid-1986, SFCL was the only foreign-invested joint venture. There were 22 licensing agreements with foreign companies, with some factories having more than one license. Four factories had co-production agreements. Each project was a part of SIC's five-year plan. SIC coordinated the funding plan and also provided financial guarantees. SIC also had an import-export department that helped in negotiations with foreign companies. Most factories had a high degree of vertical integration, but some were "supplier" factories and only produced component parts. As a result, a lot of "sourcing" took place among some subsidiaries. Usually, the SIC subsidiaries could deal directly with one another, but if there were pricing disagreements or delivery conflicts, for example, SIC could play the role of arbitrator. Sometimes SIC subsidized a factory to make its price "competitive."

Individual factory incentives had been introduced recently. Allocations were made jointly by SIC and the Shanghai Bureau of Finance. All factory profits were submitted to the Finance Bureau, taxes were deducted, and then allocations were made for building and factory development; for collective welfare, for example, health and schools; and for bonuses and other incentives. According to Zhu Ruifen, the bonus pool for each factory was decided by the SIC and the Bureau of Finance. The factory's profits were a major factor.

The factory director had some discretion in the distribution of incentives the factory received from the bonus pool. The process of implementation was described as being "very complex." The ceiling on individual bonuses and compensation was set by the Bureau of Finance. Sometimes special "extras" for model workers could be paid directly by the government. Incentive policies for the factory director were decided by the SIC in an assembly of representatives from factory workers and of engineers.

Zhu Ruifen, however, pointed out that SFCL, being a joint venture, operated differently. Incentives were decided by the Board of Directors, which was relatively autonomous. SFCL's Board chairman was SIC's GM, so SFCL had a direct link with SIC. Zhu Ruifen emphasized that SFCL was, and is, a very important project and therefore very capable Chinese personnel had been assigned to it.

NEGOTIATIONS

The negotiations between MMBI and Foxboro started in 1979 but were not concluded until 1982. In the opinion of Zhang Songlin of the Instrumentation Industry Bureau of MMBI, the long delay occurred because of the general lack of experience with China's new Joint Venture Law. As a result, all issues had to be discussed in detail.[2] Zang Zhenfeng, of the Shanghai Instrumentation and Electronics Import and Export Corporation, commented that U.S. companies, in general, that come to China initially have very limited knowledge about Chinese equipment and technology. Therefore, initial proposals are not satisfactory and the companies sometimes propose manufacturing equipment that the Chinese already have. He recommended that U.S. companies in general should contact relevant Chinese corporations and set up preliminary meetings and visits. This was a general point, not necessarily referring to Foxboro.

According to Zhang Songlin, in the initial stages of the negotiations, Foxboro was interested in supplying assembly technology for kits, whereas the Chinese side wanted design and manufacturing technology. In addition, Foxboro was reluctant to invest capital without knowing the payback period. So China invited another U.S. company to discuss setting up a joint venture. However, Foxboro agreed to provide more technology and investment. According to Zhang, Foxboro executives did not know in the beginning how many customers the joint venture would attract. However, after seeing data from Chinese sources about the large number of potential customers, Foxboro became more interested.

According to Wang Minghui, senior engineer of SIC's Import and Export Department, it was very difficult to "price" technology transfer because there really was no open market. Each company's technology was unique, and although it might be comparable to some other technologies, it had its own distinct characteristics. Therefore, each technology transfer project had to be priced individually.

Zhang suggested that U.S. companies apply to the U.S. government for approval to export the technology to China before even beginning negotiations. In his view, such an approach would save time in the actual export of technology after completion of negotiations. In Foxboro's case, for example, Chinese com-

---

[2]The Joint Venture Law was enacted in 1979, but implementing regulations were only published in 1984.

panies had already imported Foxboro products before initiating negotiations. In view of the demonstrated interest on the Chinese side, Foxboro, in Zhang's opinion, could have verified the U.S. government's willingness to license export of technology and expedite subsequent approval. The negotiations took three and a half years in all, and the joint venture received its business license from the Chinese government in the fourth year. In that year, Zhang commented, Foxboro only shipped a lot of documentation but no hardware.

START-UP

Wang Dawei, SFCL's chief accountant, stated that the Chinese government had two objectives in terms of its joint ventures: (1) to export and earn foreign exchange, the exclusive objective of some joint ventures, such as hotels, and (2) to upgrade technology. SFCL focused primarily on the second objective.

The SFCL contract was signed in April 1982, and preparations for start-up commenced in September 1982. At that time SFCL sent 20 trainees to Foxboro for training in management. SFCL adapted Foxboro's management systems in the technical area relatively easily because SFCL produced to the same quality standards as Foxboro maintained internationally. SFCL also adapted management systems in other areas, following the initial training, to facilitate communication with Foxboro. Systems in finance and accounting, however, could not be adapted readily because China had its own tax laws. Also the accounting system was different because China had a law regarding accounting systems for joint ventures. Therefore, SFCL developed reports in two different formats—one for Foxboro and one for the Chinese government; the latter went to the Tax Bureau, to SIC, and to SIEB, the overseeing authority.

According to Wang Dawei, the success of SFCL rested on two factors. The first was the cooperation between Foxboro and the relevant Chinese ministries and agencies. The second was the cooperation betweeen the GM and the DGM (one was a U.S. expatriate and one a Chinese). Each person focused on what he did best. For instance, the U.S. DGM or GM would not try to do something that required dealing through Chinese channels; rather, he would ask the Chinese manager to handle such matters.

SFCL's chief engineer (a U.S. expatriate) was responsible for the implementation in terms of selection of trainees, equipment, tools, and integration with existing processes. In his experience, for the U.S. training to be effective, the trainers had to be informed about the machines, equipment, and infrastructural conditions in China so that the training would be appropriate. In his opinion, the only problem with training in the United States was that of language and communication. Foxboro also had a training base in Singapore, a very convenient location. In addition, it was much cheaper.

The documentation that SFCL received from Foxboro filled a big room at the plant. After receipt, the documentation had to be adapted due to the different projections and standards in use in China. It required a lot of effort, even though most of SFCL's engineers understood English.

MANAGEMENT OF THE ONGOING PROCESS

Foreign Exchange

Zhang of MMBI commented that Foxboro helped in developing local suppliers. However, many components were still purchased from the United States. In his opinion, if the joint venture was to survive over the long term, the foreign exchange problem must be resolved. To do this, Foxboro needed to develop sources in China for components, even for its U.S. plant. In addition, Zhang suggested that SFCL should be the majority investor in a new hotel in China in order to generate foreign exchange required by SFCL.

The original concept for achieving foreign exchange balance involved three components:

- □ SFCL would manufacture the product for sale to the Chinese customers, except for computers, which would be supplied by Foxboro, USA.
- □ Also, component parts available in China would be purchased in China.
- □ Sales and service functions would be provided by SFCL even for direct sales by Foxboro in China. The service was in fact being provided by SFCL at the time.

According to Zhang, the agreement with Foxboro was entered into on these assumptions. All three methods of generating foreign exchange were being pursued but on a more limited scale than expected. In 1986, less and less was purchased directly from Foxboro by Chinese customers and therefore the activity (service for direct sales from the United States) had decreased. According to Wang Minghui of SIC, Foxboro was very strict in qualifying local components and therefore localization was proceeding more slowly than planned.

Chinese customers could pay SFCL in renminbi or U.S. dollars; the price was higher if in renminbi. The Shanghai government passed a policy allowing joint ventures to receive foreign exchange from customers that had foreign exchange allocations. Therefore, for a customer to pay in U.S. dollars, the customer needed a foreign exchange allocation. In that sense, each customer contract was unique. In Wang Minghui's opinion, customers with foreign exchange probably got higher priority. Wang also expressed disappointment that Foxboro had recently discontinued production of a Spec 200 product line in the United States and had transferred its manufacture to the U.K. instead of to China. He commented that if the product were manufactured in China and exported, it would help SFCL improve its foreign exchange situation.

According to Wang Dawei, foreign exchange was a problem. SFCL's primary objective was to transfer technology into China. However, the big gap between the U.S. and Chinese level of technology resulted in a scarcity of local components and inputs. Charles Rathke, SFCL's chief engineer, was chairman of the Localization Committee. He commented that he had to push local vendors to upgrade their technology. Some vendors were subordinate to SIEB, and Rathke attempted to influence them through SIEB. He used the example of a resistor component that had been qualified by SFCL. This resistor component was of a high grade and was made by one of the SIEB factories at SFCL's request. SFCL, however, only needed 200,000 units, which was a relatively small quantity. Therefore, SIEB asked SIC to ask all its instrumentation factories to purchase the component, thus increasing the demand to two million units, which made it worthwhile to manufacture. As a result, SFCL had, in effect, upgraded the component quality for the entire instrumentation industry in Shanghai.

Relationship with Local Bureau

According to Zang Zhenfeng of the Shanghai Instrumentation and Electronics Import and Export Corporation, Chinese law specified that a government organization must be responsible for managing every business unit. SFCL fell under SIEB which passed on information on government regulations and policy changes to SFCL. SIEB also did arm's-length monitoring, but most of the policies were made by SFCL's Board. In the beginning, the Board had a member from SIEB. In mid-1986, the member was replaced because, according to the Joint Venture Law, SIEB could not have a representative on the Board.

According to Zang Zhenfeng, SFCL came to the SIEB for support in the following areas:

☐ Finding local sources of components. First, SFCL contacted factories directly to purchase components. Then, if necessary, SFCL asked the SIEB to assign factories to make products according to U.S. military specifications
☐ Meeting foreign exchange problems
☐ Getting tax exemption on imported components

Mr. Zang described how the SIEB helped SFCL in 1985 to select factories capable of producing components to U.S. military specifications. Chinese factories usually produced to commercial standards, which were not as stringent as military specifications. Another problem existed: Foxboro required small quantities of very high quality components, which factories could not produce in a cost-effective way. Therefore, the SIEB sometimes had to subsidize factories for producing components.

Technology Assimilation

Zang Zhenfeng described the stages in the development of world instrumentation technology as follows:

Stage 1: Many separate instruments

Stage 2: Late 1960s and 1970s, integrated instruments

Stage 3: Late 1970s and early 1980s, modules and PC boards

Stage 4: Early 1980s to present, microprocessor technology

In his opinion, SFCL in 1986 was in stage 3 and moving to stage 4. In 1982, when SFCL was set up, SIEB, in Zang's opinion, thought that SFCL's technology was good. He believed that SFCL quality was good and the products represented a technology in wide use.

With regard to the question of whether SIEB transferred learning from SFCL to other companies in the Chinese instrumentation industry, Zang commented that it did not happen with other companies that made the same products as SFCL. However, SFCL worked with other suppliers and other systems and components manufacturers, so that indirectly they also received expertise and technology. He commented that the Chinese goal was to learn from many different sources and to develop a system of their own. They were finding that every U.S. company was distinct and had its own special contribution to make.

A big difference between SFCL and other Chinese factories, according to Zang, was that the latter were completely controlled by the state. These factories had to submit detailed plans, get approval from the municipal bureau, and sometimes from the central government. Even their sales were limited by their plans. But SFCL was controlled by its Board, which had discretion in sales, organization of departments, and so forth. SFCL used computers to support management functions but the other factories' organizations were not suitable for such support. Because these domestic factories were so different from SFCL, very limited transfer of management techniques could occur to these domestic companies.

FUTURE PLANS

The demand for SFCL's products did not appear to be a constraint. SFCL's prices were higher than those of other domestic products, but they were lower than those of imported products. Also SFCL's products could be paid for partly in renminbi. The rate of expansion, however, depended on success in maintaining the company's foreign exchange balance.

CASE 3

Westinghouse Electric Corporation: The U.S. Perspective

INTRODUCTION

On September 9, 1980, Westinghouse signed a 15-year licensing agreement with the Ministry of Machine Building Industries (MMBI) of the People's Republic of China (PRC). This contract became effective on February 4, 1981. The purpose of the agreement was to transfer technology for 300- and 600-megawatt thermal power plant turbine generators to three provinces in China—Shanghai, Heilongjiang, and Sichuan. In Shanghai, the technology recipients were the Shanghai Turbine Works and the Shanghai Electric Machinery Manufacturing Works. In Harbin, the capital of Heilongjiang, the affiliates were the Harbin Turbine Works and the Harbin Generator Works. All these production units were subordinate to MMBI. In Sichuan, the designated affiliate was Dong Fang Electrical Equipment Company. The U.S. Export-Import Bank (EXIMBK) financed the technology transfer with a five-year loan for $28.7 million. This was the first EXIMBK loan to China, since the Chinese government had regarded the EXIMBK interest rate as too high.

Since concluding the contract, Westinghouse has provided the Chinese with continued support and consultation, thereby increasing the likelihood that Westinghouse will make sales in China in the years ahead. In 1986, discussions surrounding other licensing agreements and joint ventures were in progress.

This case presents a study of the processes that led up to the conclusion of the 1980 licensing agreement, as well as the ongoing interactions among Westinghouse executives and the Chinese recipients. The case is based upon a series of in-depth interviews conducted with key Westinghouse personnel involved in the power-generation technology transfer to China. In addition, this case was based on company records and other archival data about the licensing agreement.

The Westinghouse Electric Corporation

Westinghouse, called "Circle W" by its employees, historically had been a leader in designing and manufacturing a wide variety of products including power-generation equipment. From the founding of the company by George Westinghouse and production of the first AC generator in 1900, to the first commercial nuclear steam turbine generator produced in America in 1957, Westinghouse has consistently developed leading edge turbine-generator technology.

After 1957, much of Westinghouse's scientific development in its 21 decentralized business units was dedicated to a wide variety of products. Westinghouse continued to explore new technologies and manufactured equipment in 190 manufacturing plants and 86 service and repair plants in 21 countries as of 1984. Ninety of these plants were located outside the United States. In addition, Westinghouse operated approximately 1,000 field service and distribution organizations. The company's product lines included a myriad of electronic and electrical equipment and services for electric utilities, industrial and construction market applications, and electronic systems for defense and broadcasting services. Its businesses ranged from broadcasting and cable television operations, community development, bottling and distribution of beverage products to transport refrigeration and financing services.

In 1986 Westinghouse was organized into four operating groups—the Energy and Advanced Technology Group, the Industries and International Group, the Commercial Group, and Westinghouse Broadcasting. In addition there was a corporate staff organization and a Westinghouse Credit Corporation that provided commercial financing services. Policy-making responsibility in the company was carried out by the Management Committee, which consisted of the chairman and vice chairman, the group presidents, the chairman of the Broadcasting and Cable Company, the senior executive vice president of Finance, and the senior executive vice president of Corporate Resources. Specifically, the Energy and Advanced Technology Group provided research, development, production, and support services for high technology defense equipment, such as radar, aircraft electrical systems, communications systems, and marine propulsion and launching apparatus. The Group also designed, developed, manufactured, and serviced nuclear energy systems, conventional power-generating equipment, robotics, factory automation systems, and defense systems. The Industries and International Group's activities included the design, manufacture, and marketing of products for distribution, transmission, and control of electrical power. The Commercial Group consisted of businesses having distinct customers, marketing and distribution channels and technologies.

Finally, Westinghouse Broadcasting owned and operated 5 television stations, 12 radio stations, and a production company.

Within each group, Westinghouse was organized into strategic business units of which there were 21 in 1986. These business units were free-standing businesses with responsibility for their own marketing, strategic planning, personnel management, customer service, and profit and cash flows.

Like other old-line manufacturing companies, Westinghouse has had to completely restructure a number of operations and abandon many traditional businesses in the face of constricted markets and intense global competition (Wall Street Journal, 1986). Westinghouse's second largest operating group—Industries and International, which manufactured everything from safety switches to power transformers—posted consecutive operating losses in 1983 and 1984, after earning 25 percent of the company's total operating profit in 1981. In 1985, that group reemerged into the black; however, its earnings represented less than 2 percent of its estimated $3.78 billion in sales. In that year, three of the four major operating groups posted higher sales and profits compared with 1984, with the exception of Industries and International. Total net income for the company for 1985 increased to $605.3 million, from $535.9 million for 1984 and $449 million for 1983. Westinghouse's goal was to establish itself firmly in growing markets abroad, despite intense competition. Efforts were made to expand business in the Pacific Basin, particularly in Korea and the PRC.

In 1985 over 26 percent of the corporation's total revenues came from international sources. Countries that had extensive Westinghouse manufacturing operations included Australia, Brazil, Canada, Germany, Ireland, Saudi Arabia, Spain, United Kingdom, and Venezuela. In 1985 the corporation had licensing agreements with 130 companies in 42 countries. The corporation's International Organization was responsible for coordinating the overseas activities of domestic Westinghouse businesses (exports, technology transfer, licensing, joint venture arrangements, etc.), as well as the management of Westinghouse subsidiaries and manufacturing facilities in other countries. The country manager worked with the U.S.-based product line and field sales management to direct business functions in the country. Thus, Westinghouse was in a position to implement plans that were consistent with the aspirations, requirements, and constraints of each country. The International Organization administered management and human resource development as well as helped to coordinate planning, marketing, treasury, accounting, law, and communications worldwide.

One of the central strategies employed by Westinghouse's Douglas Danforth, chairman and chief executive officer, was to put the company back in the "winner's circle" of U.S. corporations by strengthening its stance in the international marketplace.

While international business was down generally during 1983 and 1984, a number of operations performed very well, including the Asia-Pacific region.

Notable in this region were the affiliations with the People's Republic of China and Australia.

DEVELOPMENT

China in Westinghouse's World Strategy

Westinghouse had targeted the Asia-Pacific region in the late 1970s and early 1980s with a series of strategic investments in such countries as the People's Republic of China, Taiwan, the Republic of Korea, and Southeast Asia. Overall, Douglas Danforth and the management organization sought to propel Westinghouse ahead aggressively, and, thus, China was seen as a potentially critical component in their world strategy.

The PRC was therefore seriously investigated initially by Eugene Cattabiani, the executive vice president of Power Generation, and eventually by Robert Murphy, then an engineer working out of the Philadelphia office. Robert Murphy, as director of China Business Development within Power Generation, was assigned to investigate China with the support of a small task force; he was also temporarily relieved of other activities. Penetration of the Chinese market at that time was important to Westinghouse because it held the promise of a large, untapped market for a variety of commercial and manufacturing projects including thermal power plant turbines and generators.

In power generation, China was to become, over the next five years, *the* market, contracting for more power generation than the rest of the world together. Since the Chinese considered power generation to be critical and basic to its continued growth, Westinghouse perceived that, by licensing its technology to China, it would realize an opportunity to increase sales while simultaneously assisting the Chinese to attain their goals of power and energy development. Westinghouse's role in China began after normalization of relations between the United States and China in December 1978.

A theoretically ideal match was conceptualized in late 1978, with Westinghouse seeking entry into the China market and China needing sophisticated thermal power plant turbines and generators. China's "energy sector was so important [according to analysts], both as political symbol and as economic resource, that the success or failure . . . in this one sector, by whatever yardstick the results may eventually come to be measured, may well determine the fate of the Open Door more generally" (Ho and Huenemann, 1984, p. 140).

Early Contacts

Dick Gaskins, corporate director of China Programs, now assigned to the Beijing office, noted that Westinghouse's initial contacts with China in the post–World War II period came between 1945 and 1949, just prior to the foundation of the People's Republic of China. During that time, approximately 80

Chinese trainees had actually gone to Westinghouse plants for up to two years of technical and managerial training. Some Chinese engineers received Master's degrees and others obtained advanced education in the United States without qualifying for a graduate degree. Some of these people were in responsible positions in China at the time the relationship was restored in 1979, and they helped to further Westinghouse's position.

Dick Gaskins noted, "The friendships and loyalties carry over very strongly. You know, it's Chinese tradition—once a friend, always a friend, unless you let them down badly."

Following President Nixon's 1972 visit to China, then Westinghouse chairman and chief executive officer Donald Burnham; two other Westinghouse executives, Eugene Cattabiani and Lynn Saunders, director for Power Systems in the Far East; and a delegation of other U.S. businesspeople went to China in 1974 under the auspices of the National Council for U.S.-China Trade. When Burnham returned to the United States, he decided that Westinghouse should initiate discussions with the Chinese. Between 1974 and 1979, Lynn Saunders, Bob Murphy, and a few other Westinghouse executives made several trips to China to explore mutual cooperation. However, it was not until 1979 that Westinghouse was invited by the Chinese at their initiative to reestablish contact. The message was transmitted by the Chinese embassy in Washington, D.C., to the Westinghouse Washington office. A visit to China was suggested by Wang Xi Ye, a deputy minister of the Ministry of Machine Building Industry (MMBI), to explore mutual cooperation in the transfer of power-generation technology. At this time (November 1979) the Chinese indicated that they would entertain a specific proposal from Westinghouse. The Chinese wanted to establish a position of strength in fossil fuel power-generation, and it was their hope to eventually develop nuclear power capability. A delegation visited China in November 1979.

In the late 1970s, the Chinese government agency that is now the Ministry of Water Resources and Electrical Power (MWREP) recognized that the steam turbine generators produced in Chinese factories subordinate to MMBI were outmoded and were not competitive on the world market. Thus, in 1979, the Chinese decided to go to world competition and to seek technology transfer for 300- and 600-megawatt units. At that time, the Chinese were producing units of 60-, 120-, 200-, and 300-megawatt capacity. It was reported by several Westinghouse executives that the Chinese had at one time attempted to build a 600-megawatt unit, but none were ever installed; presumably they had experienced problems.

Clovis F. Obermesser, president of Asia Pacific, discussed the background of Westinghouse's licensing agreement with China in the following way:

> What we have done is transfer technology to China. We . . . made a commitment in 1979 that we would be involved with China. That was a strategic commitment. That, China being what it is, and having opened its doors to the West, our having

been involved in China in the past was important in a number of ways. We saw a corporate need to get involved quite extensively again in China.

However, the real commitment was made in 1979, after the delegation led by vice chairman Douglas Danforth went to China to discuss the ground rules for the proposal. Danforth was accompanied by 40 Westinghouse executives, and they were received by an equally large contingent of Chinese, a "significant portion of China's hierarchy," as it seemed to Obermesser. The Westinghouse executives conducted seminars in Beijing and Shanghai, principally to expose the Chinese to the Westinghouse product lines and systems capabilities. This visit was the key introduction that sounded Westinghouse's commitment to China.

Competition in Turbine Generators

In 1979, the Chinese invited all of the companies who had basic technology in fossil 300- and 600-megawatt turbine generators to submit proposals. The competitors included Westinghouse, General Electric, Brown Bovarie from Switzerland, Alsthome-Atlantique of France, Siemens and KWU of Germany. The Westinghouse executives speculated that the Chinese were not particularly serious about the two German companies, since in 1980 they narrowed the field to the other four companies.

One Westinghouse executive noted that Chinese factories had been working with primarily U.S.S.R. technology in Harbin and primarily Skoda/Czechoslovakian technology in Shanghai (probably a combination of Western and U.S.S.R. technology). He speculated that the technology in both cities was most likely equivalent to that of the United States or Western Europe in the 1950s. The Chinese intended to build the 600-megawatt unit in Harbin and the 300-megawatt unit in Shanghai.

The competition in 1980 for winning this licensing agreement was stiff. George Dubrasky, a management representative of Westinghouse's Power Generation group, resident in Hong Kong, remarked:

> If you look at world market share historically, GE has the higher world market share. They're bigger, have more licensees, and it seems to be characteristic of the Chinese that the first thing they look for is *who* has the world market share, *who* is the leader. They want to buy from that leader. So, in that case, we came in second place; probably a strong second, but in second place.

In the 600- and 300-megawatt units, Westinghouse as well as G.E. technology was circa 1960, whereas the French technology was slightly more modern. The reason offered was that neither Westinghouse nor G.E. had had any substantial market share in fossil-fuel turbines and generators since 1973 and, accordingly, had not had any incentive to update their designs.

Political Environment and Its Changes

In addition to the competitive environment of the late 1970s and early 1980s, the political environment had been changing to reflect the Chinese economic reforms. One of the most important aspects of economic modernization was the opening of China's doors to foreign trade and investment after a 23-year freeze. The open-door policy was adopted in 1976, and with it came a radical change in the political environment. Francis Kao, a physicist and director of training for China Projects, and one of the early Westinghouse negotiators, noted that the Chinese appeared hesitant to open up to international trade because of the experiences of exploitation they have had. Nevertheless, in the late 1970s, change appeared to be underway in China.

Bob Winston, director of International Trade Policy, stated that China's economy is expected to experience substantial growth in the near future and that U.S. companies would be well advised to participate:

> If you looked at a longer term view of what happened in Japan . . . most companies recognize that more effort and more resources should have been applied to getting in on the ground floor in dealing with the Japanese. The Asia-Pacific [region] has the fastest growth of any part of the world now. China is coping with its problems, and I think it has the potential to have the success story of Japan. China should grow at least at the rate of the Asia-Pacific Basin in the next 20 to 30 years.

Westinghouse executives commented that in terms of overall political changes that have occurred, among the most notable were what they labeled the Chinese ''aggressiveness'' in wanting to do business with the United States. There appeared to be at that time an appreciation of America and American things. The Americans felt that, in general, the Chinese were ''business-like in their behavior'' as well as responsive to Westinghouse's corporate commitment and strategy. And they were hopeful that all the recent political changes were in the right direction.

Two political considerations were crystallized by the Westinghouse executives. They felt that U.S. businesses thinking about entering the China market should investigate the possibilities on a *macro-political* level, as well as a *micro-political* level. Macro-political consideration focused on the stability of the political regime; micro-politics considered the operating level, for which relevant information could be gathered from U.S. personnel with China experience. At the macro level, Westinghouse judged that China represented a reasonably stable country where the risks of entry were reasonably well understood and judged not to be excessive. At the micro level, Westinghouse also felt that the company had a good chance of becoming a technology supplier to the Chinese. It was believed that such entry in the early 1980s into the China market would lead to further investment opportunities at a later point.

NEGOTIATIONS

"Catch-Up" Strategy

The personnel involved in the Westinghouse-China venture varied depending upon the stage of the negotiation process. However, when Vice Chairman Douglas Danforth led the important 1979 delegation, Westinghouse's commitment to China was established. Although Danforth made several subsequent trips to China, the primary responsibility for negotiations in turbine generators passed to Robert Murphy, director of China Business Development and the lead negotiator, and Lynn Saunders, director of China Corporate Offices. Francis Kao, also part of the early negotiation team, was the only Chinese-American involved in early negotiations, and he served as language monitor and occasionally as translator of highly technical material.

In January 1980, Robert Murphy, Eugene Cattabiani, and several other Westinghouse negotiators were invited to China to submit a bid on a contract for the licensing of power-generation equipment. According to Robert Murphy, "the Chinese really never intended to invite us to participate in the bidding. They gave Danforth and Cattabiani an invitation only to be polite, not thinking Westinghouse would pursue it." Apparently, the Chinese were engaging in serious negotiations with G.E., Alsthome-Atlantique, and Brown Bovarie.

In the January 1980 meeting, Bob Murphy "pressed the Chinese hard" to give Westinghouse the opportunity to demonstrate that the company did indeed have the "capability and wherewithall to do all the necessary catching up." Murphy said, "Just give us the opportunity," and they did.

> They invited us to come back the following month, in February, with all of our necessary specialists and so forth, and at that time we held discussions in Beijing and went to the factories in Harbin and Shanghai to analyze the factory needs. We went over in February and March, did an analysis of their needs, and then had another meeting in Beijing. Again, they were putting us through the test. Finally, they said, "Now that you know what we need, you *must* have your proposal in to us in ten days time." They said this on March 7, 1980.

Bob Murphy knew that they did not have the time to return to the United States and develop a proposal. He called for a proposal production team from Pittsburgh and Philadelphia to convene in Hong Kong. The Westinghouse team produced a proposal "crude by U.S. standards, but OK by Chinese standards," only to be unable to get air space back into Beijing by March 17. Murphy recalled the frustrations associated with getting back into China.

By chance, however, an affiliate of Jardine Matheson, Westinghouse's representative in Hong Kong, was just then trying to start up a charter air service into China. Murphy and a few others flew from Hong Kong to Beijing on the very first charter flight. They delivered the proposal to the Chinese on March 17, and in Murphy's opinion, "it really impressed them." The Chinese told

Murphy that they needed "several days" to study the material. Within three days, the Chinese told Murphy, "You are caught up." That put Westinghouse squarely in the competition. At that point, Murphy was told that "they had to study the Westinghouse proposal, and they would call back in May." Murphy recalled that the Chinese "called us back the second week of June." He claimed that Westinghouse knew it was going to be a long negotiation, but the elapsed time was longer than expected.

Contract negotiations began in June 1980. Bob Murphy noted that the negotiations with the Chinese were notably different from those with other countries: "The Chinese put all four competitors into the same building, referred to as the *Negotiations Building*. G.E. negotiators were in one room, Alsthome negotiators were in one room, Brown Bovarie negotiators were in one room, and we were in one room." There were key Chinese negotiators who negotiated with each foreign company, and a few overseers of the four who floated around and monitored all negotiations. Murphy commented, "It's a tough setup for negotiations. You can see all the hopefuls going back and forth."

At times during the negotiation process, only one executive from Westinghouse was present, usually Bob Murphy or Lynn Saunders. However, that one executive had the authority to make decisions on behalf of Westinghouse. It was this support that Westinghouse granted its senior negotiators that proved critical during the negotiation process. The schedule of negotiations included (1) the technical aspects of the technology, (2) a series of general technical parameters, and (3) the commercial and financial aspects. All U.S. personnel came directly from the Westinghouse operating groups or corporate staff and were chosen principally because of their expertise in functional areas. Translators were provided by the Chinese side.

The Westinghouse group struggled with the decision of whether to have a representative firm assist them in discussions on entering China. In 1980, Westinghouse had an agreement with Jardine Matheson for representation in Hong Kong, and the latter had already established a Beijing office. Murphy anticipated that Jardine's could be helpful, but it dealt primarily in fields distant from Westinghouse's interests. In March 1980, Murphy discussed the arrangement with Jardine Matheson, but Murphy believed that Jardine's demands were unjustifiably high; however, Murphy agreed to a liaison relationship on a monthly basis. Over time, this liaison relationship proved very helpful, principally by helping to smooth out communications.

Negotiations Venue

From the outset of negotiations in 1979, it became clear to Bob Murphy and the other Westinghouse executives that the only appropriate venue for commercial contract negotiations was on Chinese turf: Chinese negotiators had to frequently consult their superiors since they had limited authority to make on the spot decisions. The American negotiators, on the other hand, had broader latitude in negotiations. Between June and late September 1980, with the exception of a five-day period, Bob Murphy was engaged in negotiations on-site in China. During that five-day period, Lynn Saunders took over.

Bob Winston actually went so far as to say, "I don't think they'll ever let themselves get caught negotiating outside of China. . . . The problems would be too great." Indeed, in the late 1970s and early 1980s, the Chinese were going to the United States to study and to learn, but virtually never to negotiate. During negotiations on this licensing agreeement, Westinghouse did all the negotiating in China. Murphy said that he preferred to have at least one other U.S. negotiator there with him at all times. He felt the team he put together initially was the right team, and despite the fact that some had to be "rotated out of China" for rest periods, Murphy believed that the group worked well together.

Different Negotiating Styles

The Chinese concept of time influenced many of the negotiations. The Chinese had a 6,000-year history, and ten years to them was simply a blip on the curve. George Butterfield, director of China Operations, said, "I don't think anybody really, really appreciates that. We're so impatient! If we wait two years, we think we've been the most patient people in the world. They're still after things today that they wanted four years ago; . . . they're very patient. If I say no now, they believe they'll get it later."

There were considerable differences in negotiating styles related to time utilization. The Americans tended to make relatively quick decisions, whereas the Chinese were considerably slower. John Denman suggested that, with respect to time, "the first thing you do when you arrive is that you tell them when you will leave. Everything is geared to that time. They also want to know what the agenda is for the time that you're going to be there." Dick Gaskins told us:

> They have gotten nicer, though. I remember in the early 1980s, occasionally your airplane reservations were mysteriously cancelled if things hadn't happened. That happened with Bob Murphy on two occasions. . . . You would be over there negotiating, and you'd tell them you were leaving at such and such a time, and your reservations evaporated . . . they were gone!

Other stylistic differences included risk taking and information processing. Dick Gaskins discussed risk taking, by way of comparing the risk taking of Americans with that of the Chinese:

> I think the biggest impediment to negotiations in China is the fact that the people you are negotiating with are not risk takers. They are extremely reluctant to take risks and therefore they can't make a decision while they're sitting at a table. Whereas our people are basically compensated on their ability to take risks . . . acceptable risks. So, when you have two people, or two Westerners negotiating, there's a give and a take that occurs during the process itself. Whereas the Chinese can't do that. You've got to tell them what you want. They've got to think about

it, and then they'll come back and tell you whether or not they can accept it. And, if you can't accept it, you have to tell them. And then they've got to go back and think about it again. It really does extend the process quite a bit.

Further, all intermediate negotiation results and summaries of every negotiation meeting had to be committed to paper. This served as the "collective memory" of the negotiators. However, handling paperwork in China takes considerably longer than it does in the United States. All negotiations documents were written in Chinese and then typed in Chinese, which was quite a chore, since they did not have word processing. Then it had to be translated, and the English version had to be approved. U.S. negotiators usually tried to have the Chinese translate their documents into English, since Western negotiators cannot use Chinese drafts.

During negotiations, the Chinese always wanted the Americans to "show their hand" first. The Chinese typically did not prepare a Chinese draft until they had received the first one in English. One executive commented, "It is a negotiation tactic. They are more interested in knowing where you're starting from because they are afraid that if they prepare the draft, they may concede something."

Thus, key differences emerged between the Chinese and American negotiators in terms of their styles. There were, of course, some individual differences attributed to intercultural training and personal experiences, according to the U.S. negotiating personnel. The Americans in general had virtually no intercultural training in preparation for negotiations, whereas the Chinese were keen observers of business negotiations with foreigners. The American propensity to quickly "ask for the sale" was repeatedly stymied by the Chinese need to constantly check with their superiors.[1] The effective approach was for the American negotiators to tell the Chinese negotiators in advance what they wanted and then to give them a chance to return prepared.

Westinghouse negotiators saw the Chinese negotiators as being shrewd and certainly "out for a bargain," according to George Butterfield. "I think they just expect that people somehow are just going to naturally help them . . . and naturally want to do it for nothing." The Chinese negotiators typically had little authority to make binding decisions, and as mentioned, their negotiation contingent had to seek the approval of a decision-making authority, regardless of how many Chinese were present at the negotiation table. In general, the Americans had very few negotiators present during the negotiation process. The Chinese always maintained a large contingent of personnel, typically in the rear of the room, whom the Westinghouse negotiators did not know. They weren't introduced, and when asked, the Chinese usually replied that they're from "the

---

[1]After 1981, intercultural training was made available for Westinghouse executives by a contract with an outside consulting firm. Some executives claimed that Westinghouse personnel who went to China relied on published materials including Lucian Pye's *Chinese Commercial Negotiating Style* (Cambridge, Mass.: Oelgeschlager, Gunn and Hain, 1982).

Ministry." Often these participants were observed to take copious notes or to make telephone calls, according to Westinghouse personnel.

Butterfield commented, "You can't let that structure worry you too much. You might go in [to a negotiation session] yourself, and you're 1 out of 15." That is, one American facing 15 Chinese.

Finally, the Chinese propensity to keep seeking a "better deal" interfered with their ability to maintain a long-term relationship with one manufacturer. Therefore, the constant comparison of prices between manufacturers in the early stage of negotiations eventually gave way to more serious and thoughtful investigation into the Westinghouse product. One executive commented, "The Chinese do their homework well, as I'm sure you know. And, they've learned from other countries who have experience in this regard. So, they really try to get every 'i' dotted and every 't' crossed on what's going to transpire."

Bob Murphy added that time seemed unimportant to the Chinese negotiators. It did, however, appear to him that "people were important. Obviously, they spent a lot of time studying Bob Murphy and getting to know his hot buttons." When Murphy brought someone new to the negotiation table, the day was wasted, since he felt the Chinese were studying the new person.

Choice of Product to Be Transferred

The Chinese were, during the late 1970s and early 1980s, keen on receiving world-class equipment and technology that would vault them into a more modern era. In discussion of the events that led up to the technology transfer, George Dubrasky reminded us that the product that they were getting from the Chinese factories was 20 years behind world standard. Dubrasky credited Westinghouse's competitive advantage to a "willingness to be willing teachers and give them the whole thing." The "whole thing" included such intangibles as know-how and the "craft" side of the technology, not just manufacturing drawings. This paper included drawings, aperture cards, process specifications, design manuals, computer programs, computer tapes—in short, anything needed for the manufacturing as well as the design phase of the product.

According to Dubrasky, they looked at the Alsthome generator, the G.E. turbine, and the Westinghouse turbine, and they wanted all of that "mishmashed into one product." "But, you can't do that," someone replied.

John Denman, chief counsel of the International Law Department, commented on this attitude of the Chinese and explained the learning that took place that helped resolve the problem.

The Chinese kept pressing for the transfer of more technology than was covered by the agreement. There were some "tough conflicts," according to George Butterfield, over what was and was not to be transferred during the start-up phase of the affiliation. "They keep coming back and saying, 'Yeah, we should get that motor technology.' They're saying they need that in order to be self-sufficient on this thing. That's true, but somewhere in there it's an interpretation of the contract." In many cases, Westinghouse accommodated the Chinese side. Ultimately, the business unit decision to license 300- and 600-megawatt power-generation technology wasn't subject to the severe scrutiny it might have undergone had the technology been anything other than mature.

Method of Transfer

Dick Gaskins referred to three stages of technology transfer that generally apply and that were also relevant to the China project. The first was the transfer of documentation, which included relevant software programs. Second, was the training of Chinese personnel, and third, consulting. Although the technologies under discussion were advanced, they were nevertheless mature technologies that Westinghouse had been manufacturing for quite some time. A comment was made to the effect that different countries had different methods of transferring technology. Bob Winston told us, "It's not uncommon for the Germans to have a technology transfer where they transfer the manufacturing drawings only. That's like having the prints to a car; then how do you put it together?" The implication was that Westinghouse was prepared to transfer both the documentation and the know-how to manufacture the equipment.

One problem that arose was how much documentation should be supplied. Gaskins commented that one way "is to take the approach of transferring the now-existing technology frozen at a certain point. Any new developments are subject to renegotiation." This would be static technology transfer. Carl Hamner described an alternate dynamic approach:

> It has generally been our tradition to license a product and to give the licensee any improvements to that product that Westinghouse puts into commercial production during the period of the license. So that if you have a . . . 300-megawatt steam turbine, model number such and such, and there are additional improvements that are made to that during the term of the license agreement, Westinghouse generally has included those as part of the package.

Westinghouse tended to favor dynamic licensing agreements. Dick Gaskins commented that the static approach "guarantees constant dispute," because any subsequent modifications would require additional money to be paid by the Chinese. During the negotiations Westinghouse executives felt that the Chinese needed repeated reassurances that Westinghouse intended to give them "everything" and that a dynamic process would respond more to their perceived needs. Westinghouse supplied, as part of the package, what they referred to as trade secrets, patented know-how, as well as unpatented know-how.

A second problem that arose was that of "information overload" on the part of the Chinese, which would have happened had they been supplied with all expected development and manufacturing drawings and changes for any of the licensed products for the entire time of its production. Some of these products had been in production for decades and therefore had multiple refinements to the original product. On the other hand, the Chinese expected to obtain all of this material. Bob Winston commented that this was resolved by saying to the Chinese, "We'll give you the drawings that will enable you to make what we are selling." This position was eventually acceptable to the Chinese side.

A third problem encountered was that Westinghouse was purchasing some of the components used in the manufacture of turbines from outside suppliers and therefore did not own that technology. Westinghouse could only contract for the transfer of technology that it owned directly. Understandably, all Westinghouse could transfer were specifications of externally sourced components they used and the source they bought it from, but not any manufacturing or design information.

The problem was made even more complex by reorganization within Westinghouse. The Power Generation Division had to pay other divisions to get technology that previously had been under their control. The Chinese believed that if it had the Westinghouse logo, they were entitled to the technology regardless of whether it was outside the purview of the Power Generation Division.

This problem was partially remedied by negotiating various categories of information that would be transferred. The first was a category of information over which the Westinghouse Power Generation Division had total control. In this case, in addition to manufacturing data, they provided limited design information. The second category was information on components bought from "noncorporate" companies. In such cases Westinghouse supplied material only describing the purpose of the specifications in addition to the manufacturing data.

This pointed out a critical difference in perception between the Chinese and Western perspectives concerning any given technology. George Butterfield commented as follows:

> Up front, [the Chinese] think the technology on turbine generators consists of a whole lot of paper, exactly how to. In other words, like you might put together a recipe to make a cake . . . a well-defined package that you ship over. I'm sure that they believe that we're going to give them a list. A list of all the things that they [need to build a] 300-megawatt turbine generator: You put this, this, and this in and you have a cake. Now, the only added key to that would be that in [the recipe] there would also be the ingredients for making flour. That's just not the way Western technology really is. In other words, the interrelationships between components and how they function are important ingredients of technology.

The difference in perception on the scope of a technology added a layer of complexity and delayed the start-up process.

Joint Venture or License

Very early in the interaction process, a question arose concerning whether Westinghouse should engage in a joint venture or a licensing agreement. Prior to the Joint Venture Law and even just after its enactment in 1979, the context in which one could establish a joint venture was unknown.[2] Bob Winston added, "If we had approached them on a joint venture, they probably would have thrown up their hands and said, 'We'll deal with someone else.' " Thus, Westinghouse tried to introduce the idea of licensing without overlooking the possibility of a future joint venture. Obermesser suggested that if they redid the agreement today, they might have negotiated a different type of agreement. Clearly, Westinghouse negotiators were searching for an agreement that would satisfy both sides at a minimum, which ultimately meant profitability. The lack of an experience base in both licensing and joint ventures caused these executives some concern. We were also reminded that Westinghouse had provisions for joint venture conversion in some of its technology transfer agreements with companies in other countries.

Contracts

The licensing agreement was concluded on September 9, 1980, at 9:00 p.m., with the formal signing of the American and the Chinese contract. It was a significant time in that the Chinese liked the symmetry of the 9-9-9. Dick Gaskins affirmed the use of contracts retrospectively, despite repeated attempts at "renegotiations" and uncertainties concerning the legal basis for enforcing the contract.

> In the event of a controversy over the interpretation of the contracts, I don't know the extent to which you will be able to enforce those rights in China. The only thing you can really rely on is that the Chinese are very concerned about the face that they present to the world. And that if you have a serious dispute with the Chinese on a contractual issue and it is resolved in an impartial forum, and the Chinese refuse to acknowledge that, they will lose a great deal of credibility. They care about that. If they abandon that concern, then I would say that the contract probably isn't worth the paper it's written on.

The licensing contract was subject to both governments' confirmation. From the U.S. perspective, this meant obtaining an export license, which was issued on February 4, 1981.

START-UP

Choice of Chinese Factories

Westinghouse had no input whatsoever as to which Chinese factories were to be the recipients of the 300- and 600-megawatt turbine generator technology. They were asked by the Chinese officials of MMBI to visit four factories—a turbine factory and generator factory in both Harbin and Shanghai. The Chinese

---

[2]Implementing regulations to the Joint Venture Law were not published until 1984.

officials asked Westinghouse to scan the facilities, note the existing machine tools, and come up with a proposal for a layout of machine tools that would be necessary to manufacture the 300- and 600-megawatt verification units.

Bob Murphy discussed the Westinghouse executives' observations by noting the mammoth sizes of the factories. Each factory had between 6,000 and 8,000 employees and was spread over hundreds of acres. Murphy claimed that all four factories seemed equally capable.

A third factory site, Dong Fang, in Sichuan Province, was designated to receive boiler feedpump drive turbines. This type of technology, according to Murphy, was far lesser in scope, and the Dong Fang factory was dissatisfied with the implied lower status.

The Harbin factories were designated to produce the 600-megawatt verification unit, to be installed in Ping Wei, in Anhui Province. The Shanghai factories were charged with completing the 300-megawatt verification unit, to be installed in Shi Heng, in Shandong Province. Dong Fang, under the purview of Canadian Westinghouse, was charged with manufacturing boiler feedpump turbines.[3] Murphy suggested that the start-up process for Canadian Westinghouse was not well integrated with what was occurring in Harbin and Shanghai. Murphy claimed that their relationship with Dong Fang has always been somewhat unclear.

Paper and People

As discussed, "paper and people" were the critical elements in the turbine generator technology transfer process. In terms of paper and people, Westinghouse attempted to define the package that the Chinese partners wanted. However, this procedure was very different from those used with other Westinghouse licensees, according to Bob Murphy. When companies of most other nationalities became licensees of Westinghouse technology, it was "across the board, from A to Z. Here the Chinese only wanted two segments—the 300- and 600-megawatt class units." Westinghouse spent a considerable amount of time using its best engineers in coming up with a suitable design. Initially, China requested a "national design" for both class units, a generic design to "knock out turbine generators with the ease of a cookie cutter." They ran into a lot of problems there and are not pursuing that track today.

With regard to the start-up procedures, there were several stages of implementation that began in 1982. The first stage entailed moving drawings and information on what is known as contract products to the Chinese affiliate. During the second stage there was much training and consulting, bringing people

---

[3]Shortly after the 1980 signing of contracts, Westinghouse transferred boiler feedpump turbine technology to Canadian Westinghouse. Thus, even though Westinghouse had the responsibility for implementing the technology transfer, they lacked the responsibility to negotiate parts and component sales for that technology. Further, they did not have the authority to direct Canadian Westinghouse to do something, as the Canadian affiliate was a free agent.

over to the United States or Westinghouse people going to China. This process, which started after the export license was granted and the contract was officially approved in 1981, continued well into 1985. The third stage included the engineering design and development associated with the 300- and 600-megawatt units.

Dissemination of documentation. Westinghouse described assembly and transfer of the documentation for the 300- and 600-megawatt turbine generators as problematic. The state of the documentation varied considerably because, as with any mature product, there was a lot of craft that had evolved over time. This presented a problem during start-up because there had "never been the need" to maintain sophisticated documentation on the "know-how" aspects of product manufacture.

Dick Gaskins commented that this was "a very, very sophisticated and extensive technology. It's literally thousands of drawings, shelves and shelves of manufacturing information, and so on." He added that the bulk of the documentation that was transferred to China was "literally railroad cars full." One Westinghouse employee calculated that about 16 tons of paper were sent. Westinghouse was unwilling to commit their R&D laboratories in Churchill to providing all the new developments over the past 15 years in the form of paper, unless the Chinese were willing to purchase the new developments at additional expense. However, Murphy noted that if in the design of the transferred technology Westinghouse found a structural weakness, they owed it to the Chinese licensees to inform them. If, on the other hand, further developments improved the design and performance of the transferred technology, the Chinese would have to pay for that documentation. As of 1987, the Chinese had been unwilling to pay for the latter class of developmental improvements to the 300- and 600-megawatt class units, despite the fact that numerous improvements have been made.

With respect to the paper that had been thus far transferred, Murphy noted that there would be "dribs and drabs up until the twelfth month of the fifteenth year." They had disseminated approximately 95 percent of the paper thus far. The remaining 5 percent would most likely concern developments to the product. He commented that the Chinese side checked carefully if any drawings were missing.

Dissemination of paper was also achieved by holding technical seminars. Technical seminars were held once a year, designed to inform the Chinese as to what Westinghouse was doing in state-of-the-art technology, not necessarily specific to this technology transfer. This gave the Chinese a flavor of the new technology that Westinghouse was innovating without having to pay for the information.

Training and consultation. Executives at Westinghouse identified training as a critical and continuing aspect of the start-up process. At the corporate level, executives listed training and the foreign exchange problem as major hurdles to be overcome in this licensing agreement. Training was identified as pivotal in the successful transfer of technology, since, in one executive's words, "the training really teaches the Chinese how to go from the drawings to the hardware." Training transfers the know-how upon which the manufacture of the product depends. In addition, training included advice on and supervision of modification of the Chinese factory to make it suitable for product manufacture. The Chinese had extended the training window significantly, hoping to avail themselves of newer technical developments in that way. Westinghouse has attempted to narrow the window. Bob Murphy commented, "They can absorb a technology that is nonexistent today if they do their training tomorrow."

The 15-year license agreement specified three stages of training. The first stage was "total A-to-Z training on product, including product design, product development, quality assurance, manufacturing engineering, as well as management and accounting control training." Westinghouse voluntarily gave them marketing training, which alone amounted to approximately 60 man-hours.

The second stage, referred to as systems design training, was subcontracted to Charles T. Main. Of the 2,000 man-months of overall contracted training, 200 man-months were designated for training in systems design.

The third stage was a repeat of the first stage. According to Bob Murphy, "Their concept back in 1980 was that they were going to send over their upper-level and mid-level management people for stage one, and their younger engineers for stage two. But what happened, in fact, was that stage one and stage two went into a blender and became a mish-mash." As of today, seven-eighths of the 2,000 man-months of training have been utilized. Westinghouse did not perceive the need for additional training as of 1987 and felt that whatever training credit remained could easily be transferred to consultation.

The training that accompanied these phases was scheduled to begin in October 1982 and to conclude in June 1984. The training, however, ran behind schedule and it had not been completed as of 1987. Sometimes the Chinese scheduled three people to come for training, but it happened that one didn't come. The Chinese wanted Westinghouse to repeat the same module for the missing person. It was a major task to coordinate the time schedule for the training. So far, Westinghouse had not refused any of these requests. In addition to training in the United States, there were also training sessions later held in China. The training in the United States was carried out in modules and typically in a classroom setting. The training site was generally Philadelphia, a unionized plant that posed very few problems, according to Murphy.

Gaskins elaborated on the Chinese recognition of the importance of the type of training to teach the Chinese not only to review and read drawings but also "what Westinghouse technology is." The initial turbine generator contract called for 2,000 man-months of training. The Chinese had since expanded that by 10 percent to 2,200 man-months of training. It was Westinghouse's belief that they were thinking about additional training over the 2,200 figure.

Francis Kao commented that this training was equivalent, in number of hours, to sending more than 500 people through 15,000 weeks of training. He added that one training problem was the interpretation of these man-months of training. Kao stated that the Chinese always wanted to send more people for less time. He stated, "They want to send four people over here for two months instead of two people for four months." The Westinghouse executives believed that the Chinese propensity to send greater numbers of people for shorter periods of time allowed more trainees to be sent to the United States. Therefore, more Chinese would be exposed to this type of turbine technology. In addition, being sent abroad was considered a prize by most trainees.

Kao emphasized, however, that language remained the major problem in training, subsequently affecting the efficiency of the training process. The Chinese sent interpreters with the trainees; however, these interpreters generally lacked adequate language skills for both technical and general matters. Not only did the language problem affect the efficiency and effectiveness of the learning process, it created unnecessary misunderstandings. Kao added that he monitored the training modules for communication problems, and when he sensed or felt there was a language problem, he alerted the trainers.

Kao made several suggestions to improve the communication process in training sessions held in the United States and those subsequently held in China. These suggestions included the following: (1) slowing down the speed of the instructor's speech and putting as much as possible in writing; and (2) monitoring body language, for example, watching for "questions in people's eyes" and for "sudden discussion" among the students, both signs of communication problems.

Some Westinghouse executives noted the general lack of adequate language skills in China, referring to the lack of adequate comprehension of spoken English. However, Kao observed that the engineers that attended the Westinghouse training sessions were very capable in reading English technical material, having read considerable Westinghouse technical literature.

Some Westinghouse personnel voiced frustration concerning the shift of trained personnel to management positions. Murphy noted that "once they were trained, they almost never went back to utilize that training. They became too valuable from a management standpoint." He further stated that "a majority of them aren't working in the area in which they were trained."

Consulting

Consultation by Westinghouse personnel in China was also part of the second phase of the start-up process and was perhaps the most controversial. There was a major difference between the parties to the agreement concerning what the term *consultation* entailed. To Westinghouse, consulting meant helping project management to keep going; whereas to the Chinese it meant sending an expert for a few days whose sole purpose was to teach one particular area of expertise.

Bob Murphy noted that approximately 350 man-months of consultation were to have occurred in the United States and 50 in China. Westinghouse needed to place a cost on them. Murphy commented, "I chose to define a man-month as the number of their people who spent time in our facilities. That was their people, not ours. When Westinghouse personnel went to China to consult, it was the number of our people who spent time in their facilities." Westinghouse generally accommodated the Chinese demands with respect to consultation because, as George Dubrasky put it, "The consultation part of this [technology transfer] . . . [has to be given] to them or they'll never be able to build it. And they make some mistakes when they do things the first time. But they go back and do it over again rather quickly."

A basic discrepancy in the Chinese and Westinghouse consulting perspective was based on payment for knowledge. Dubrasky noted that the Chinese resisted paying for expertise and would frequently request a consultant to help them. What took 25 minutes to do in China would cost Westinghouse $10,000 to send over a consultant. Another problem arose when the Westinghouse consultant had to do a lot of background work, just to spend only a few hours actually consulting. Butterfield stated that the Chinese would have liked to have a full-time Westinghouse consultant in their plant. He noted that Westinghouse sent a consultant to assist the Shanghai Generator Works for a period of three months at Westinghouse's expense when the latter considered this assistance essential for the quality of the product.

While Westinghouse may have had 12 to 15 people in China as consultants at any given time, they were there only on a part-time, temporary basis, with the one exception noted above. The Chinese desire for an on-site, full-time consultant led to an attempt to recruit outside consultants for project management. Dubrasky commented, "They [the Chinese] have approached [Westinghouse] retirees on an individual basis, but they have not concluded a deal. I don't know whether they could. . . ." It was implied that the Chinese agencies could not come to satisfactory agreements on remuneration with the retired Westinghouse personnel whom they approached.

MANAGEMENT OF THE ONGOING PROCESS

Overview

By 1985, the paper transfer associated with the start-up had been completed between Westinghouse and the Chinese affiliates, but the training and the consulting processes were still continuing. Prototypes of the turbine generators were completed by 1984; however, they were still in the testing stage in 1985. None had gone into manufacture, nor had any been installed in China as of 1985. While the Westinghouse consensus at that time was that the transfer was going well, it would take time before the Chinese affiliates were capable of sustaining continual wholly-independent manufacture of turbine generators.

Renegotiations

Attempts at renegotiation of the contract signed in 1980 by the Chinese began later that year. George Butterfield suggested a strong dichotomy between the Chinese and the Westerners in their adherence to the legal contract. It was quite common for the Chinese to reopen discussion on items contained in the contract documentation. It was unclear to many of the Westinghouse executives just what the impact of refusing their repeated requests for renegotiation would have on long-term relations. It was also noted that individuals, to make themselves look good, frequently reopen negotiations in an attempt to make a "better" deal subsequently. Since state-of-the-art technology was critical to the Chinese, it was understandable that they continued to press for renegotiation for widening the scope of the agreement to include additional items.

Communication

Successful management of the ongoing process relied on communication and feedback between China and Westinghouse. Communications problems included the lack of a permanent, resident Westinghouse employee in China; the Chinese practice of selectively communicating or avoiding "casting criticism on themselves"; and the apparent lack of lateral communications between Chinese Ministries (presumably MMBI and MWREP). George Butterfield commented that "the communications aren't really good." Absent a full-time resident, the data Westinghouse received were minimal. On the other hand, the Chinese willingly communicated information on Westinghouse wrongdoings. George Butterfield observed that Westinghouse executives always heard about what they were doing wrong and only infrequently and indirectly what the Chinese were doing wrong. Obermesser highlighted the communications difficulty in getting a "line of approval" for any action:

> I think their ministeral classifications are still extremely private. There seems to be a lack of relationship between one Ministry and the other.[4] Therefore, to get a line of approval, you're never sure you've covered all the bases. So there's a very frustrating [spectrum] of situations out there. But, by the same token, patience pays off. And, somehow, at some point in the chain, we are really able to identify what it is that we must do to go on with the business we carry on out there.

---

[4]The ministries in this case were the Ministry of Machine Building (MMBI) and the Ministry of Water Resources and Electrical Power (MWREP). The MMBI was the unit with which Westinghouse negotiated the licensing agreement; they had direct authority over the three manufacturing sites. The Ministry of Water Resources and Electrical Power was the power utility and customer of the manufacturing plants (or of the MMBI, which was the supplier).

Manufacturing Planning and Schedules

As of 1983, it was apparent that the technology transfer process was behind schedule. Westinghouse and its Chinese counterparts expected that it would take 60 months for the 300-megawatt unit and 72 months for the 600-megawatt unit to be completed. The expectation was that at the end of the third year, both products would be in production. As mentioned above, the Chinese were only at the prototype testing stage as of 1985. This delay was partly due to the tremendous amount of information involved. It simply had not been possible to be absorbed in such a short period.

Performance

Because production was in the prototype stage, performance quantity had not become an issue. However, as Gaskins pointed out, quality was a vital component of performance. "Their performance does impact the way they eat [or their quality of life], so they're going to get a lot more [quality oriented, as time goes on]. It's in their self-interest."

In any licensing agreement the licensor relinquishes control over the quality of the final product. Obermesser stated that when you license the technology, you "lose the quality aspect of the technology by passing it to someone who doesn't know how to use it. But I think what you try to do in transferring technology and what the licensee tries to acquire in getting your technology is also the quality aspect, or he won't be interested."

All of the executives agreed that in a licensing agreement such as this, the Westinghouse name was not put on the product and therefore quality was not, strictly speaking, a paramount issue for the licensor. However, in possible future joint ventures it would be, but then Westinghouse would have more control.

Bob Murphy noted that the Chinese were hesitant to share production goals with Westinghouse; however, he was able to ascertain that by late 1987 or 1988, the Harbin factory planned to have the production capacity for turbine generators of total capacity of 4 gigawatts (GW) or 4000 megawatts (MW). Surprisingly, he noted that they only anticipated going up to 6 gigawatts of total power-generation capacity by year 2000, even though their physical plant could turn out 15 gigawatts, in Murphy's opinion. Also, Murphy believed the Shanghai factory is slightly ahead of the Harbin factory.

Local Content and Sourcing

Variability in metals and electronic components somewhat slowed the production process, since some components had to be developed through local production and sourcing and others had to be imported. Bob Murphy noted that in 1984 Westinghouse knew that China's development of electrical power-generation capacity was seriously lagging. Murphy called a meeting with the two other companies, Combustion Engineering and Ebasco, which were the suppliers of boiler and systems technologies, respectively, and which had been selected to work with Westinghouse on the verification units. Murphy suggested the formation of a consortium that would cooperate to meet the intermediate requirements of the domestic power-generation industry. It was suggested that each U.S. firm survey the Chinese plants with which it was affiliated to determine the plants' production capabilities during the near and long term.

Also in 1984, the Ministry of Water Resources and Electrical Power (MWREP) petitioned for, and received central government approval, to import a power plant of 10-gigawatt capacity rather than develop the resources internally. As it happened, the local factories underrepresented their production capacity to the MWREP due to politics between MWREP and MMBI. The MWREP had previously said that they would buy all the factory capacity for the next several years. Since the factories appeared unable to meet the demand, at that point, MWREP was free to request permission from higher authorities (the State Planning Commission and the State Economic Commission) to go outside China for importation of additional capacity. In this sense, Bob Murphy claimed that "both the factories and the MMBI were mousetrapped."

To implement the import of the 10-gigawatt power plant, the central government formed a new corporation called the Hua Neng International Power Development Corporation (HIPDC) in February 1985. Since the Westinghouse–Combustion Engineering–Ebasco consortium was already operational by that time, it participated in the bidding on various contracts conducted by HIPDC. Murphy claimed that "the three of us were really totally committed to building the electric power industry in China. We hired Wharton to do an independent study on the value of using local content, and we went to the factories to determine quotations as to what they could produce within our scheduled requirements." In August, after a closed bidding process, the consortium was informed that it was not on the short list of HIPDC.

Murphy said,

> They [the consortium] were prepared to talk to the Chinese on price, but they didn't give us the opportunity. In the meantime, we went a step further; we went over their heads. We went to the State Economic Commission, we went to the State Planning Commission, and we went to the Bank of China, and emphasized that they had to look at the long-term picture. But the HIPDC's charter was to import, not to build local industry. They wanted to have an all-imported plant, because they could have a greater degree of control that way, even though each consortium member was taking full responsibility for product guarantees and performance guarantees. By and large, our consortium was discriminated against because of our local content. At that time, we told HIPDC that we could include approximately 27 percent of local content.

"The higher authorities' position was that," according to Murphy, "you have an interesting story, and we wish you hadn't told us all this." Murphy's thought was that "we were embarrassing everyone. We were an outside group interfering in internal politics, and it wasn't appreciated, particularly by the customer, HIPDC."

That bidding was followed by another bidding in which Westinghouse again used local content and again lost the contract, this time to Toshiba at less than one-half a percent price difference. Toshiba had not previously transferred any power-generation technology to China. Bob Murphy noted with some dismay that the Chinese will tell you that the use of local content is an internal goal to foster self-sufficiency, yet incidents such as the above continued to regularly occur. It was Westinghouse's position that, whenever possible, they will endeavor to utilize Chinese products subject to restrictions imposed by the Buy-American Act and EXIMBK financing.

Assessment of 1985 Position

Despite the problems outlined above, Westinghouse in 1985 was reasonably satisfied with the progress it had made in China. This technology transfer was the first step in working toward a long-term business involvement in China. The second step would have been a joint venture. China must have also been content; in 1985 it invited Westinghouse back for other discussions around licensing thermal power plant technology. Clearly, Westinghouse had achieved a corporate presence in China; it had people who worked there, lived there for periods of several months, and followed up on problems that required great effort. Bob Winston exclaimed, "If you swept up all hours spent by the people throughout the company who have an involvement in China, you could easily end up with 15 man-years."

By March 1986, the Chinese had built prototypes of the steam turbine generators and were testing them. However, at this point, none had been installed. George Butterfield commented,

> Getting the technology took a lot longer than [we and they] expected. The Chinese expectation was that they would be manufacturing turbines in three years. So it would have been something like 12 to 18 months up front for assimilation of the technology, the drawings, the manufacturing processes and then they were immediately planning on jumping into a normal schedule.

Two reasons why the project took so much longer than expected were, according to Westinghouse, (1) the Chinese inclination to wait until all of the documentation had been transferred before proceeding with the manufacturing processes, and (2) the fact that project management was not included in the license agreement. One executive commented,

> One of the things in the six years is kind of a two-way street, their method of waiting until the technology was all there before they started on the next step. The second thing that I think took six years was [to develop appropriate] project

management [which] was not part of the contract. The agreement was for us, Westinghouse, to provide the technology, to provide the training, to provide parts and components essentially. That's what we got paid for.

FUTURE PLANS

Overview

China is expected to play a lesser role in Westinghouse's overall strategy than might be expected of a nation in excess of one billion people. On the risk side, Westinghouse eventually recognized that China was not a high grade investment market. The major difference between China and Westinghouse's other licensees was that China wanted only a small segment of Westinghouse's product line, whereas most other licensees negotiated broader licenses. All of the future interactions that Westinghouse was contemplating with China did not represent a significant risk to their overall portfolio. Therefore, if Westinghouse was not to be successful in China, its business position would not be endangered.

Westinghouse had been encouraged by its activities to date in China and was forging ahead in its pursuit of other technology transfer and joint venture agreements with China. Nevertheless, it had to demonstrate that future events were expected to occur as a direct result of its current investment of people and time. Continued investment in an economy such as China's can be of little significance to a company unless it results in further returns on investments. After five years of extensive work and effort, Westinghouse executives generally felt that they were reaching this point. As with most technology transfers to developing countries, it usually takes a minimum of three years, and more likely five years, for the company to see a return on its initial investment. Westinghouse was now entering its sixth year and yet, profits seemed elusive.

Westinghouse has rearticulated its commitment to the Chinese, and it is being rewarded by the Chinese interest in a future Westinghouse presence in China.

Future Ventures in China

Dick Gaskins explained that Westinghouse was not in the consumer market; rather, it was in the industrial market. As one executive commented, "We're not trying to sell refrigerators; we don't make consumer products any longer, so we aren't looking for cheap manufacturing. We are looking for basic, stand-alone joint ventures in the Chinese market and in the global market [with] foreign currency."

By 1985 no joint venture agreements had been concluded, yet as of 1986 Westinghouse had a number of joint ventures under negotiation. Dick Gaskins added that Westinghouse had been pushing a joint venture quite strongly, but he noted that "joint ventures are inherently slow, more difficult, and less likely to be successful."

Two obstacles to Westinghouse's deep penetration in the Chinese market included the anticipated problems with quality of products that would ultimately bear the Westinghouse trademark and Westinghouse's ability to develop more accurate management control and budgeting procedures within the constraints of the contract. An issue related to management control that Westinghouse must consider in all future agreements concerned the need for more accurate forecasting of costs and budgeting. A breakdown of all items over the course of the license should also be included in such forecasts.

A third obstacle to Westinghouse's deep penetration of China, which was less likely to be under the direct control of Westinghouse and more difficult to predict, concerned increased competitive market factors in China. One executive noted that Westinghouse and other major manufacturers of turbine generators were in a buyers' market, since good turbine generator technology was available from a number of suppliers. Obermesser suggested,

> Competitive? Cripes, it's probably the most competitive arena in the world, because they never stop getting competitive information. And once they have it all, they still don't have it all. It is a continual flow of tough negotiations and, as I say, probably one of the most difficult, competitive places that we have dealt with in the recent past.

Carl Hamner agreed, "The carrots there are so large that every major competitor, for the most part, is engaged in some kind of activity."

The Chinese political situation so far had been very favorable for Westinghouse and other companies. Westinghouse executives acknowledge the pivotal role that politics in both countries played in their future in China. Westinghouse's long-range presence in China was inexorably linked to that country's plans for the future, plans that had been known to shift in the past. China's continued modernization depended on a market economy, material incentives, more encouragement for entrepreneurs, and increases in foreign investment, all of which had to be conducted within a socialist environment. It was a delicate balance, which, if maintained, could result in a prosperous relationship for both China and Westinghouse.

Epilogue

In August 1987, the 300-megawatt verification unit had been successfully tested and had gone on-line in Shi Heng, Shandong Province. The second 300-megawatt unit had not gone into production, yet Westinghouse was informed that it would be operational by August 1988. The 600-megawatt verification unit was completing its shop testing and was to be operational in Ping Wei, Anhui Province, during the second quarter of 1988. Westinghouse was not aware at that time of plans to begin manufacturing the second 600-megawatt unit. The 1980 contract stipulated a six-year payback schedule; and in 1987, the

Chinese were delinquent in their foreign exchange payments. The power industry appeared to be in political disarray.

Robert Murphy and George Butterfield were posed the question, "If you had it to do over again, what would you do differently?" Both wrestled with the question and, in the end, both agreed that they would most likely take the same general approach in view of the customer's stated goals. From a project management standpoint, however, both agreed that there would be a lot of changes. Murphy expressed his opinion in the following way:

> You're not in there to sell something and walk away. You want to be a participant in the long-term growth of China, and this is what we have been striving for. We saw the technology transfers as an entry with the next step being joint venture. Through this joint venture, we would take them into the international world to export Chinese products. Three or four years ago when we first presented these ideas to the Chinese, they expressed interest, but more recently I feel that they have weighed the value of outside interference in the management of their factories versus the potential increase in productivity, and decided it's not in their interest to have outside interference. They're taking the most comfortable road, and in China, there are no risk takers.

Murphy also went on to say,

> In the early years of [China's] new government, it has been so drilled into them that they must have self-dependence. They are a generation away from being able to get away from that mentality. Most factory directors spend more time determining whether this guy should marry this woman or whether this person's mother is getting the proper medical care, than he does running a factory or producing a quality product. How can he produce a product when he's got this responsibility of running a 8,000-member community? So ultimately, you can lead a horse to water but you can't make him drink. We have talked very openly with our Chinese associates and told them what our 1980 objectives were. The Chinese said to us then that we needed to help them now by giving them the technology, and then a whole lot will be open to us in the future. So far, that hasn't happened.

Murphy described a series of power plants that went to contractors other than Westinghouse, including GIE of Italy, Alsthome-Atlantique, Toshiba, and BMW. Murphy implied that Westinghouse was on its last measure at this time. He questioned Westinghouse's continued presence in China if the hoped-for joint venture was unattainable. He continued to pledge Westinghouse's commitment to the Chinese, and he emphasized the concept of participating with the Chinese. Within weeks he should be informed of their decision. Without a joint venture forthcoming, Murphy thought that Westinghouse would diminish its emphasis on the China market. However, Murphy stressed that they were not going to give up on China until "the whole door is shut." As a company, they were still seeking the right combinations and assumed that if a firm doesn't have an operational base in China, then "you're a short-term participant in the world. One in five people on this earth is Chinese."

Nevertheless, at this writing, it was anticipated that the new Westinghouse chairman, John Marous, will deemphasize Europe, the Middle East, and the Far East in favor of domestic business. Since the international market was currently below cost level, Murphy explained that while an international presence cannot be "turned off and on" easily, one needed to consider the long term.

That long term will be affected by, among other things, political factors outside the control of U.S. firms. By way of explaining Westinghouse's lack of control over the technology transfer process, Murphy explained that Westinghouse's future was caught up in political factions in China that fought constantly. In short, "the right hand doesn't talk to the left hand, or else they compete with one another, usually at the expense of the client."

Murphy surmised that the Westinghouse reputation has been affected since the Chinese perceived that the power plants should have been on-line "a long time ago." Another example of past problems haunting Westinghouse, explained Murphy, was that the Chinese did not tell them anything.

> The factories won't talk to us; the central group in Beijing won't talk to us about what they're doing as a follow-up. They're not even telling us what the obstacles are. Combustion Engineering is also going through the same kind of frustration. So there's a certain amount of not only frustration, but disenchantment, with our relationship with the Chinese.
>
> One of the things that has been so difficult to deal with is that we're not dealing with China prior to 1949. We're dealing with China today, 1987, and even though we never, never talk politics there, politics are there. Friendship now is, "What can you do for me today?" It's been so hard to realize that there's that type of political change, and this is bound to affect all U.S. businesspeople.

To summarize, Westinghouse has been disappointed by recent actions taken within the power-generation equipment sector in China and has concluded that the market does not hold out sufficient incentives for further efforts. At the same time, Westinghouse is pursuing activity in other sectors of the industrial equipment market in China.

EXHIBIT 1   Westinghouse Electric Company: Key Events

Date	Event
1945–1949	80 Chinese trainees sent to Westinghouse for technical training
1974	Chairman Donald Burnham's visit to China; Burnham returns to the United States and encourages Westinghouse to proceed in investigating the China market
1978	U.S.-China normalization of relations, open-door policy begins
1978	China goes to world competition for technology transfer of 300- and 600-megawatt class units of fossil power plant technology
1979	
November	Vice Chairman Douglas Danforth leads the Westinghouse contingent to China; Westinghouse invited to participate in the competition after a late entry
1980	
January	Preliminary negotiations begin in earnest; Robert Murphy is lead negotiator
February	Westinghouse visits the factories in Harbin and Shanghai to analyze factory needs
March 7	The Chinese tell Westinghouse it has caught up with the other firms in competition for the bid; Westinghouse is told that a formal proposal is due in ten days; Murphy calls in a proposal team to Hong Kong, and within ten days Westinghouse produces a proposal
March 17	The proposal delivered to the Chinese
	Bob Murphy establishes a liaison relationship with Jardine Matheson for representation in China
Mid-June	Contract negotiations begin
September 9	15-year Licensing Agreement between MMBI and Westinghouse is signed at 9:00 p.m.; 300- and 600-megawatt power plant technology bound for three provinces: Shanghai, Sichuan, and Heilongjiang
	Westinghouse opens a Beijing office and staffs a Shanghai office
1981	
February 4	Westinghouse receives its export license; the U.S. government approves the contract
1982	Beijing office is registered; start-up begins; documentation sent to China; training commences in the United States for Chinese trainees; consulting commences
1983	Chinese begin discussions with Westinghouse on nuclear turbines; their first nuclear project, referred to as 728 Project, is a 300-megawatt reactor of indigenous design; they want a co-engineering effort on a 600-megawatt nuclear reactor through a co-development effort; technology transfer is behind schedule, due to administrative burden in assimilating documentation, training, and manufacturing delays

EXHIBIT 1 (*continued*)

Date	Event
1984	Prototypes of the turbine generators continue to be in process; a Westinghouse–Combustion Engineering–Ebasco Consortium formed to assist the Chinese with power-generation needs
1985	Paper transfer of documentation associated with start-up completed; prototypes begin testing stage; first discussions occur around possible joint ventures on other power-generation projects
1986	Future ventures continue to be discussed
1987	Dick Gaskins transferred to Beijing
May 30	300-megawatt verification unit goes on-line in Shi Heng, having completed the "100 hour test"
June	600-megawatt verification unit is completing stop testing; it will be shipped to Ping Wei the second quarter of 1988; Westinghouse
Summer	executives express increasing disenchantment with China's political environment

EXHIBIT 2 Westinghouse Electric Company: Organization Chart

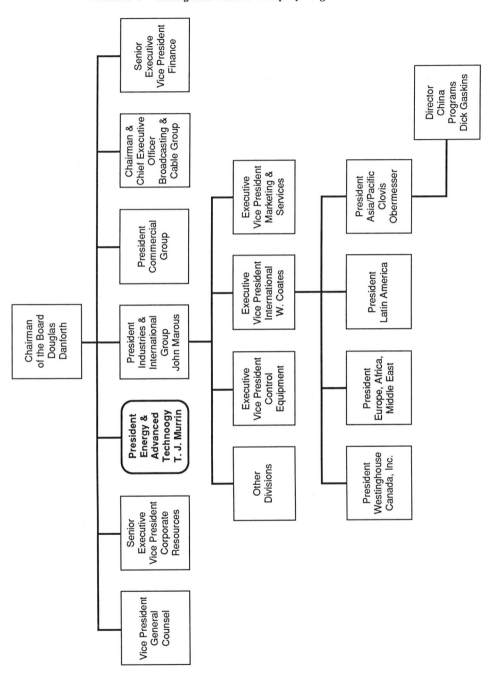

EXHIBIT 3   Westinghouse Electric Company: Energy and Advanced Technical Group

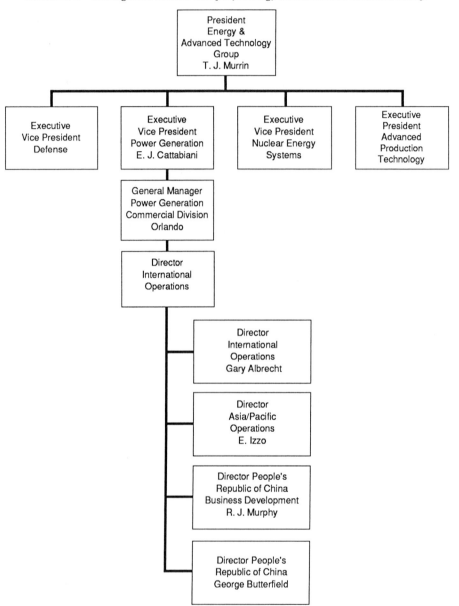

CASE 4

Westinghouse Electric Corporation: The Chinese Perspective

INTRODUCTION

The material presented here is based on historical documentation and on interviews held in China from May through June 1986. The research team visited the Ministry of Machine Building Industry (MMBI) in Beijing, the Harbin Turbine Works, the Harbin Electrical Machinery Works, the Harbin Boiler Works, and the Harbin Power System Engineering and Research Institute, all located in Harbin, Heilongjiang Province.[1] In Shanghai, we met with representatives of the Shanghai Turbine Works, the Shanghai Electrical Machinery Works, the Shanghai Boiler Works, and the Shanghai Power Plant Equipment Research Institute. These various groups allowed us to investigate the Chinese perspective of the licensing agreement between Westinghouse Electric Corporation and the MMBI, signed in 1980 and effective on February 4, 1981.[2] This 15-year licensing agreement was for the technology transfer of 300- and 600-megawatt generators to three provinces in China: Shanghai (Shanghai Province), the recipient of the 300-megawatt power plant technology; Harbin (Heilongjiang Province), the recipient of the 600-megawatt power plant technology, and Dong Fang (Sichuan Province), which was to have received 300-megawatt technology.

---

[1] In December 1986, the MMBI was merged with the Ministry of Ordnance Industry, forming the State Commission of Machinery Industry. We will continue to use MMBI in this case analysis.

[2] The U.S. export license for the technology to be transferred was issued on February 4, 1981, according to Westinghouse sources.

Discussions were held with MMBI officials who had helped conclude the licensing agreement and with the heads of the Harbin and Shanghai factories and research institutes mentioned above.

DEVELOPMENT

MMBI officials and others indicated that the relationship between Westinghouse and China began a long time ago. Their first recollection was that Westinghouse transferred technology to China in 1946, and the Chinese individual in charge of that technology transfer was Yun Zhen, now 85 years old.

The MMBI officials commented that U.S.-China technology transfer had a number of problems, largely due to differences in social systems, infrastructure, and culture. Mr. Sung told us that "technology transfer is complicated. The technology does not have to come from the original developer of the technology." Mr. Sung and other MMBI officials noted that some technology originally developed in the United States no longer maintained the lead as time went on. Furthermore, there were cost figures and different standards that complicated technology transfer. Thus, the Chinese Ministry officials told us that there were many things that figured into the Chinese decision to go with either a primary developer or the most recent developer of the technology in question.

With respect to the 300- and 600-megawatt power generation technology, the historical background began in the 1940s, when 87 engineers were sent from China to Westinghouse to study transformers, generators, switchgears, and other technology. Most of those engineers later worked at the Harbin Electrical Machinery Works, founded in 1951. As a result of that early contact, the Harbin Electrical Machinery Works, and later the Harbin Turbine Works, designed equipment during the early 1950s using Westinghouse design plans. The Chinese, between 1953 and 1960, intensified their contacts with the USSR and assimilated Soviet technology into their designs.

Between 1960 and 1978 there was virtually no contact with foreigners, and the Harbin Electrical Machinery Works, the Harbin Turbine Works, and other Chinese enterprises involved in power generation had to rely on their own resources. In 1978, when relations were normalized with the United States, we were told that the first company to become reinvolved with the Harbin Electrical Machinery Works was Westinghouse. It was within the next two-year period that China initiated the open-door policy and encouraged the development of science and technology principally through the policy which became known as the Four Modernizations. The needs at that time were for more advanced power plant technology in several sectors in China, including sophisticated thermal turbine technology, boiler technology, and generator technology.

Before the Cultural Revolution (1950–1966), technology developed at a relatively rapid pace. In 1966, the gap between the technology in China and the rest of the world was believed to be only ten years. However, during the upheaval of the Cultural Revolution, this gap widened at a time when industrial (turbine) technology developed quickly in the rest of the world. From 1976 to 1980, this gap was judged to have widened to approximately 25 to 30 years. The Chinese believed the gap to be most evident in three particular areas: economic efficiency, safety records, and automation and reliability. During the entire period since the founding of the People's Republic in 1949 until 1979, imported technology and power-generation equipment had been purchased only from the USSR. After the ouster of the Gang of Four, some power-generation equipment was purchased from Japan on the level of 125 megawatts; however, before 1980, 95 percent of total power capacity was supplied by domestically manufactured equipment.

Once importation of power plant technology had been agreed to, the Chinese reviewed the world competitors. They sent delegations to Europe, America, and Japan to determine with whom they should enter into negotiation for the contract. They then selected a group of companies who were judged capable of transferring the desired power-generation technology and invited them to China to discuss the projects. Thus, in 1979, G.E. and Westinghouse were invited from the United States, Brown Bovari from Switzerland, Alsthome-Atlantique from France, and KWU from Germany.

NEGOTIATIONS

The Chinese began their negotiations with four finalists: Brown Bovarie, G.E., Westinghouse, and Alsthome-Atlantique. During the negotiations process, which began in 1979 and concluded with the signing of a licensing agreement in 1980, the MMBI obtained as much information about the four companies as possible. They negotiated with each company separately for three months. First, they asked the foreign company to introduce its technology in depth. After these presentations were completed, the Chinese personnel questioned the foreign representatives in detail, particularly with respect to the adaptability of their technology to production in China. The Chinese officials at the MMBI judged that from the answers given to the questions posed, they would know how good the company would be to work with. The Chinese MMBI officials felt that they could judge how well the foreign company understood the problems faced by the Chinese plants and assess their willingness and capability to adapt the technology to local production. The foreign companies were also asked to describe the financial aspects of the proposal in great detail. Finally, the Chinese officials quizzed the companies on deficiencies of the latters' technologies and known rates and causes of breakdowns.

Chinese officials commented that each company's specialists attempted to answer the questions posed to them. Apparently, if the foreign engineers were not broad enough in their knowledge to answer all questions, often the company called in specialists to address specific queries. Westinghouse brought in technical specialists for short periods, simply to provide specific answers. The officials noted that G.E. also sent a few people, but they were available for comparatively shorter periods of time.

After the initial period of negotiations, it became clear that the final contest was between G.E. and Westinghouse. Originally, most of the Chinese engineers thought that G.E. was the more advanced, but G.E. eventually lost the contract. In their decision making, the Chinese officials considered Westinghouse to be more "flexible" and more "modest." G.E. was alleged to have imposed more restrictions, particularly in disallowing future transfer of the technology to nuclear application. Westinghouse posed no such restrictions. The Chinese thought it puzzling that both G.E. and Westinghouse were from the same country, yet they had such different policies regarding nuclear applications of licensed technology. The Chinese officials believed that, in general, G.E. was less open and that it was more difficult to obtain answers to detailed questions from G.E. personnel regarding technology. Thus, Westinghouse eventually won the contract.

Later on, the Chinese heard that the individual in charge of the negotiations for G.E. was "criticized" by the G.E. administration, and that GE had subsequently reexamined its China strategy, apparently having learned from this early mistake.

During the introductory stage of the negotiations, the Chinese delegation included between 40 to 50 persons, consisting of two principal groups: government officials and technical personnel representing the Harbin and Shanghai plants and others from the research institutes. As the negotiations proceeded, the delegation size was reduced to about 10 Chinese representatives, with one leader.

The negotiations were labored for several reasons, including language barriers. We were told that at the time the open-door policy was implemented and during these early negotiations, very few Chinese could speak English. Only technical personnel who were over 50 years old could speak English, as these were the ones who had had early contact with English-speaking countries. By contrast, many young people now speak English. The negotiations were initially complicated by the fact that in many cases, the Chinese interpreters' knowledge of the language was not much better than that of the technical personnel. As agreements were reached, "memoranda of intent" were also written by Chinese specialists; these memoranda had to be translated into English. The foreigners did not provide interpreters but instead brought Chinese speakers, who, during the negotiations, monitored the translation with respect to technical accuracy.

In general, the Chinese officials believed that the main problems came from the American side during the negotiations. They stated that the primary obstacle was the wish of the U.S. company to impose terms and conditions derived from U.S. law. They felt that the model contracts simply "got in the way." We were told, by way of contrast, that EEC countries and Japan did not bring lawyers to the negotiating table, since lawyers injected inflexibility into the negotiations and inhibited their flow. In particular, the Chinese officials objected to terms based on U.S. law.

The MMBI officials told us that for all projects undertaken, they had a particular Chinese enterprise in mind from the start to be the recipient of the technology. In the present case, the designees were the Harbin Turbine Works and the Harbin Electrical Machinery Works, which received the technology for the 600-megawatt class units; the Shanghai Turbine Works and the Shanghai Electrical Machinery Works received the technology for the 300-megawatt class units. Verification units were to be produced at those two sites, and eventually production was to be extended to Dong Fang in Sichuan Province.

In China, after a contract has been signed, approval must be obtained from only one ministry—the Ministry of Foreign Economic Relations and Trade (MOFERT). By contrast, the export licensing process in the United States was judged to be an interagency process involving many agencies. The Chinese interviewees felt that the approval process was much faster in China, and the Chinese side usually had to wait for the U.S. government's approval. The Chinese complained that in the United States, the contract first went to the U.S. Department of Commerce for approval. Then it went, perhaps, to the U.S. Department of Defense and others. It was believed that the U.S. government added additional requirements to the contract that had been negotiated between the MMBI and Westinghouse. It was further commented that in China, all factors were discussed between the Chinese companies, and once agreement was reached at that level, the Ministry usually approved the contract. However, the Ministry had usually been kept well informed and influenced the negotiating position during the negotiation process.

START-UP

Paper Transfer

The licensing agreement was signed in September 1980 and approved formally February 5, 1981; the paper transfer began shortly thereafter. The Chinese officials commented that some of the early documents that were needed most were delayed badly and some documentation which was less urgent arrived first. Their experience and comments to the same effect indicated that most U.S. managements (including Westinghouse) have an insufficient work force allocated to the delivery of the technical paper required for the technology transfer.

After the documentation had been received, the Chinese officials noted that documents had to be translated, metricated, and U.S. standards changed to reflect Chinese standards. In addition, we were told that the methods of projections in engineering drawings were different from those used in the United States. The officials complained that the U.S. companies never changed the documentation to reflect the Chinese specifications before transferring the paper, noting that most EEC countries do this as a matter of course. These factors have occasionally precluded the selection of a U.S. partner. The Chinese officials also noted that some of the components which they believed should have been included in the transfer were sourced by outside suppliers and were, thus, unavailable. They believed that these circumstances were not always known during the negotiation process and that they served to complicate the technology transfer.

The technology transfer agreement included the following stipulations: Westinghouse was to transfer technology for the 300- and 600-megawatt-class turbine units. The term *class* implied an additional 10 to 15 percent capacity limit, thus extending the range to almost 700 megawatts. Further, since the design of the turbine was modular, the specific modules to be transferred were stipulated. The technical methods and the design details to be included were also specified in detail. Finally, there was a protective clause stipulating that all technology that should have been covered and was not, was thereby included. Chinese officials claimed that they have had good experience with that last clause and that both Westinghouse and the Chinese have been accommodating in this respect. By way of example, we were told that the agreement stipulated the transfer of 120 computer programs, but later a Chinese technician found that there were over 200 programs that the Chinese found useful. They requested all 200 programs and received them.

Information was subsequently exchanged between China and Westinghouse, and Westinghouse designed a complete unit based upon it. Modules were used, and the Chinese plant was required to buy some parts from Westinghouse. This procedure was standard for Westinghouse. Chinese officials stated that Combustion Engineering (another company involved in the power plant technology transfer) actually helped the Chinese co-design the boiler to Chinese specifications, and they much preferred that approach. After the completion of the entire unit, the machinery had to be installed and subsequently verified. Tests were to be conducted to determine if all specifications could be met. The first unit of each class was designed as the verification unit. The assembly of the 600-megawatt verification unit was completed in December 1986 and was then shipped to the chosen site for installation at the Ping Wei Power Plant in Anhui Province. It was to be verified in 1988. The 300-megawatt verification unit was completed in 1986 in Shanghai and was installed at Shi Heng Power Plant in Shandong Province in July 1987. It was to be verified in May 1988. The Chinese affiliates expected that they would be kept informed of all technical modifications made by Westinghouse during the next decade.

Westinghouse specified materials to be used by U.S. technology standards. Sourcing in China was seen as problematic, and the Chinese officials indicated that substitutions had to be made. Westinghouse requested a substitution report for review before giving its approval. The Chinese noted that this entailed a considerable amount of work, and fortunately Westinghouse approved most substitutions. The officials suggested that a more appropriate method would be a joint design effort with the participation of Chinese design personnel at Westinghouse and with the introduction of Chinese standards during the entire process. Westinghouse was known to have contributed to some of the joint design efforts carried out by Chinese engineers with Combustion Engineering, and this project could be used as a model.

The start-up was ultimately delayed by one year, according to Chinese officials, due to delays in documentation transfer caused by lay-offs at Westinghouse following a downturn in world market conditions. Manufacturing and completion of the verification units were also delayed as a result. The Chinese felt that they could have asked for a penalty from Westinghouse for the delay, but they felt that they had a long way to go on this project (15 years) and chose to forego a fine. There were other spillover problems due to documentation delay that the Chinese officials discussed. Westinghouse sent components to Beijing by airfreight, and the transfer to Harbin was very slow due to Chinese customs delays. The Chinese requested that Westinghouse send paper and components at a minimum of one month early to compensate for the expected delays inside China. These expected delays were, according to Chinese officials, innate to the Chinese system.

In spite of their blaming the U.S. side for many delays, the Chinese admitted that revisions to the documentation came too frequently, even during start-up. Westinghouse managers told their Chinese counterparts that the frequency of revisions caused some problems even to U.S. workers. Thus the Chinese requested a "freeze" on the revisions, and this was implemented by Westinghouse. After a period of time, the Chinese side could then decide on what revisions should be incorporated into the product.

Westinghouse further accommodated its affiliates by guaranteeing that components used by them would not be classed as outdated and would continue to be available. This problem illustrated the negative side of the dynamic technology transfer, according to the Chinese officials. They believed that people charged with implementation must be able to absorb both the technology and the rapid rate at which the new changes were introduced. They therefore judged freezes as useful devices to alleviate the pressures generated by the continual flow of documentation intrinsic to dynamic technology transfer.

Training

Once the design work had been completed, training was scheduled to begin. According to the agreement, the training was to amount to 300 man-months. Individual trainees were to be sent to U.S. Westinghouse facilities for periods ranging from three to nine months, the average duration being four and a half to five months. The detailed plan for each segment of the training was established in advance including content, time, and place. The trainees also received intensive training in English prior to their trip abroad. A significant number of Chinese engineers had completed training as of 1986. It was reported that training was not without its particular problems. For example, the Chinese officials noted that in China, design engineers did the design calculations and also drafted the plans. But in the United States, the design engineer only carried out the design calculations and did not do the drafting. The result was that when the Chinese design engineers returned from the United States after training, they had not been fully familiarized with the drawings. The net result was some confusion upon translation of some of the drawings, necessitating that the Chinese design engineers return to the United States for further training.

In addition, there were also reported to be union problems: Chinese trainees were not allowed to operate machines or tools on the factory floor. Sometimes Chinese workers were not even allowed to obtain information from the U.S. workers; the Chinese officials indicated that the union objected and instructed American workers not to cooperate in transferring skills. It seems that the union was telling American workers that they would lose their jobs to the inquisitive Chinese. The officials noted, however, that once good relationships had been established with the U.S. workers, the latter mostly cooperated despite union objections.

The Chinese managers complained that the trainees received mostly classroom training as opposed to on-the-job training, which they would have preferred. It was their opinion that union regulations at Westinghouse plants contributed to the choice of training methods.

During the training process, technology transfer was further delayed by language difficulties mentioned earlier. In many cases, interpreters were sent to accompany the trainees. The Chinese officials also told us that living conditions were hard for the trainees, as they were quartered at some distance from the Westinghouse facilities and were unable to drive a car, thus necessitating that a driver be sent from China.

Thus, the Chinese officials concluded that during the first year, and particularly related to training, there were language barriers, both in listening and in speaking. They concluded that knowledge of English was critical. They also now believed that interpreters weren't effective during training, although they were effective on purchasing missions. Most of the returning trainees were put in charge of a specialized field according to their training. In summary, with respect to training, the Chinese felt that they learned a lot, but the method had its problems.

Consulting

The Chinese concluded that consulting was the link between the training and management of the ongoing process. When manufacturing got to a certain point, Westinghouse made a practice of sending a consultant to investigate and solve problems encountered on-site in China. Most of the Westinghouse consultants sent to China were good, according to the Chinese officials. However, some were not familiar enough with the systems and couldn't answer some of the specific questions. According to the contract, seven man-months of consulting were allocated per location in China for the whole project. This turned out to be insufficient, but the limit was set by the Chinese government. The Chinese claimed that Westinghouse sent engineers to Shanghai free of charge for six months to supplement the funded program when, in their judgment, such additional help was needed. Consulting was subsequently judged to be critical to the initial start-up and subsequent management of the ongoing process. As a result, Li Peizhang, deputy chief engineer of the MMBI's Electrical Equipment Bureau, agreed to increase the government's earlier allotment from seven to nine man-months. In summary on the consulting assistance the Chinese side received, they noted that Westinghouse consultants were often too specialized and that their knowledge was not sufficiently broad to answer specific questions posed by the Chinese engineers. They claimed that these consultants often failed to follow up on unanswered questions after their return to the United States.

MANAGEMENT OF THE ONGOING PROCESS

The first 300-megawatt verification unit was completed in 1985, and the first 600-megawatt unit was to be completed in 1987. The next five-year production plans included eight to ten units of 300-megawatt capacity and three to four units of the 600-megawatt class.

We were told that both Harbin and Shanghai technical personnel found the Westinghouse technology to be satisfactory. They told us that a key point to consider in decision making regarding technology transfer was that success did not depend upon the nature of the technology alone. The Chinese asked themselves whether they, all by themselves, could digest, build upon, and further develop the technology received from Westinghouse. They felt that in summary, they were 90 percent satisfied with the Westinghouse technology as judged by these criteria. Nevertheless, the Chinese noted that, in their opinion, they will always retain the image of some "not very friendly people" who relied on the letter of the contract. The Chinese paraphrased some Westinghouse personnel as saying, "According to the contract, I've given what I should."

In general, however, the Chinese felt that no major problems had been encountered. One manager said that the problems encountered with Westinghouse were "only very specialized technical problems," that the personnel and organizational problems were only temporary, and that the importance of the technology overshadowed other lower level difficulties.

In their evaluation of the ongoing process, the Chinese suggested that there are many different opinions as to what constitutes performance; they highlighted the following criteria:

1. Whether the Chinese plant can produce the whole product from locally sourced material and components after technology transfer.

2. If the production process can be implemented in a reasonable amount of time.
3. If the cost of production is reasonable. Payment to the supplier of the technology was judged reasonable at 15 percent of the total cost of the product; 30 percent was judged "too high."

The Chinese officials also noted that there are other performance criteria. They believed that, while mostly successful, there had been too many delays. Five years had passed, and the product was not, as yet, in production. In the early stages communication with Westinghouse personnel was difficult. Later the Chinese said telex facilities were installed in the factories, and the MMBI stationed a liaison officer in the United States who kept in touch with all parties, that is, Chinese trainees at Westinghouse and the U.S. companies involved in the project. These arrangements helped overcome communications problems.

Finally, Chinese officials mentioned the problem of foreign exchange as affecting ongoing performance. They noted that both Westinghouse and Combustion Engineering suffered from a shortage of orders. They claimed that the Americans emphasized profits, whereas they, the Chinese, emphasized foreign investments in China.

Further evaluation of the ongoing process included questions on whether to produce large 600-megawatt units or use smaller units. The Chinese officials noted that cost was an important factor. The present cost of one 600-megawatt unit would pay for six smaller units of 200 megawatts each. The Westinghouse generator was overdesigned they felt and required too many U.S. materials. Chinese officials ultimately felt that the optimum product would be a jointly designed 600-megawatt unit, incorporating preferred features such as lighter design and use of more domestic materials and components.

In May 1986, officials from the Harbin and Shanghai factories met in Shanghai with officials from Westinghouse for design review of the 300-megawatt unit. For the 300-megawatt unit being produced in Shanghai, the joint design was to be carried out free of additional charge, as specified in the contract. This joint design was in progress in 1986. The Chinese officials felt that a similar tack should have been taken on jointly redesigning the 600-megawatt unit, which was not stipulated in the contract. Nevertheless, in 1987 Westinghouse complied with the Chinese request and agreed to assist in the Chinese redesign of the 600-megawatt unit. This activity was referred to as on-the-job training and was included under the training segment free of additional charges.

In assessing the use of local components, it was noted that for the first unit, that is, the verification unit, 70 percent was supplied by Westinghouse and 30 percent was of local origin. Of that 30 percent, 20 percent was assembly, and only 10 percent of the value represented locally sourced components. Over the longer term, they expected that 85 to 90 percent would be locally supplied. They noted further that some materials cannot be manufactured in China; this included some special copper alloy, some plastics (Teflon), and some insulating material. Chinese officials told us that they will continue to obtain 10 to 15 percent of the components from Westinghouse, even in the long term.

The Chinese further disclosed that as they looked back on the contract, it seemed to have been implemented smoothly. They said that both sides had profited from the technology transfer project. Westinghouse had made a financial profit, and they see Westinghouse as profiting from the continued future sale of American parts and components introduced into the Chinese market. In summation, the Chinese said that they too profited by being able to catch up with advanced technology more quickly.

FUTURE PLANS

As mentioned earlier, the future calls for an increase in jointly developed models of the 300- and the 600-megawatt units. The 15-year contract stipulates joint design exchange meetings in which Westinghouse presents its own designs and the Chinese present theirs. These jointly developed models could be produced by both sides separately, since no plans have been made for joint production as yet. Two joint design meetings have already taken place, one in Florida, the other in China.

Ultimately, Chinese officials felt that they had established friendships between the Chinese people and Westinghouse. It is now planned that the Chinese will progress from the type of licensing agreement they now have, to co-manufacturing with Westinghouse (as well as with Combustion Engineering) in order to supply units to international markets, where they will have to compete against other international companies. By way of example, Chinese officials noted that together with Westinghouse and Combustion Engineering, they planned to bid for projects in Egypt (two 600-megawatt units), financed through the World Bank, in Singapore (a 300-megawatt unit), and in Ningbo, China (a 600-megawatt unit).

Discussions were also underway with the Shanghai Electrical Works for a joint venture to be formed with Westinghouse. Clearly, the next steps in the continuing relationship with Westinghouse were aimed at developing closer relationships with an eye toward movement into co-manufacturing and ultimately joint ventures. These future steps were in the exploratory stages as of 1986.

Chinese officials saw a tremendous amount of competition between U.S. companies for a claim to the China market. They noted potential problems here when two competing foreign companies had license agreements with different firms in China. For example, Westinghouse had agreements with the Harbin and Shanghai plants, and G.E. sought to enter into an agreement with Dong Fang (near Chengdu, Sichuan Province), for the transfer of 600-megawatt turbine generator technology. Since the contract with Westinghouse included terms of confidentiality, Chinese officials were worried that, should a G.E.–Dong Fang contract be concluded, they could not guarantee that there would be no sharing of technology between the different Chinese production units.

Thus, the future plans included planning for and entering into new licensing, co-manufacturing, and joint venture relationships. The Chinese appreciated the long technology experience of Westinghouse and hoped to further capitalize upon it in the years ahead.

CASE 5

Cummins Engine Company: The U.S. Perspective

INTRODUCTION

In January 1981, Cummins Engine Company, Ltd., headquartered in Columbus, Indiana, signed a ten-year license agreement with the China National Technical Import Corporation (Techimport) and the Ministry of Machine Building Industry (MMBI) Automotive Bureau, now called China National Automotive Industries Corporation (CNAIC) for the manufacture of two lines of Cummins diesel engines at the Chongqing Automotive Engine Plant. By 1985, the Chinese licensee was producing 75 units of the NH-series engines per month, with about 30 percent local content, and had begun production of K-series engines as well.[1]

Since concluding the agreement, Cummins has continued efforts aimed at broadening the company's penetration of the China market. An agreement for the establishment of a Technical Service Center in Beijing was signed in 1984 with Techimport and CNAIC. This Center was established for the purpose of supporting Cummins' dirct sales to Chinese customers. Efforts to promote additional affiliations with Chinese automotive and other engine factories, as well as efforts at establishing additional servicing facilities in China were in progress.

---

[1] NH-series engines have a range of 160 to 475 horsepower and are used primarily in trucking and industrial applications. K-series engines have a range of 450 to 1,800 horsepower and are used in industrial applications.

This case presents a study of the processes leading to the conclusion of Cummins' first licensing agreement, of the technology transfer mechanisms, and of the continuing affiliation between Cummins and the Chongqing Automotive Engine Plant. The case is based on a series of interviews with senior executives and managers of Cummins, and it has been prepared with the company's cooperation. We gratefully acknowledge the important contributions made by Dr. Andrew Chu, the present general manager of the China Business Group, who was most cooperative and made the case study possible by organizing our visit; Thomas W. Head, vice president and general manager of Affiliated Enterprises; Charles B. Byers, vice president of International Business Development; Dennis Kelley, former director of Operations, China Business Group and at present outside consultant to Cummins; and Dennis Piper, manager and engineering liaison of the China Business Group. A great deal of valuable material for the case was also obtained from the thesis by P. Marlow, to whom the authors feel indebted (Marlow, 1985).

The Cummins Engine Company

The Cummins Engine Company, headquartered in Columbus, Indiana, was described as "the world's leading designer and producer of diesel engines."[2] In 1984 the company had 21,000 employees worldwide in 30 manufacturing, assembly, and research facilities located in nine countries. The Cummins organization chart is presented in Exhibit 1 on page 166. In the United States Cummins owned two research and engineering centers, four engine production plants, five component manufacturing plants, and three engine and components remanufacturing facilities. Further, Cummins engines were produced in three plants in the U.K., one factory in Brazil, and two plants in Mexico (one being a licensee and the other a joint venture between Cummins and the Mexican government). Cummins also had a joint venture in India and licensees in Japan, China, and South Korea, as well as component manufacturing plants in France and England.

Cummins' net sales for 1984 were $2.3 billion with net earnings of $188 million. The company was estimated to have over 60 percent of the North American heavy-duty truck diesel engine market.[3] The Cummins product line included a hundred basic engine models classified in ten engine series, with a range from 50 to 1,800 horsepower. The company literature listed the applications of its product as trucking, construction, mining, agriculture, and power generation, with the percentages of total sales in terms of number of engines sold in 1984 being 63.5 percent to North American trucking customers, 22.3 percent to North American industrial firms, and 14.2 percent to international customers.

---

[2] *Cummins at a Glance,* Cummins Engine Company, Inc., 1984 Annual Report, December 31, 1984.

[3] Ibid.

Cummins was founded before World War I by Clessie Cummins, an inventive engineer who obtained the financial backing of William G. Irwin, a banker interested in creating employment opportunities for Columbus. By 1922, the company had advanced sufficiently to hire a chief engineer to design and develop new diesel engines based on Clessie Cummins' ideas (Rowell, 1980). In spite of relatively primitive conditions, the young company developed engines superior to those of better equipped competitors as witnessed by a visiting German engineer in 1925 (Rowell, 1980). During the 1920s and 1930s, Cummins experienced slow but steady development, and during World War II the company played a prominent part in providing mobile power to the Allied Forces.

The company's international business activities dated back to 1919; overseas sales formed a substantial part of the company's small-scale business operations (Rowell, 1981). During and just after World War II, Cummins' products flowed to Europe in large numbers, first with the military and subsequently during the Marshall Plan. Eventually, Cummins established a manufacturing facility in Scotland in 1956 to supply the countries of Western Europe at a time when these countries were not able to purchase from the United States because of currency restrictions. Later, in the early 1960s, production of Cummins engines was licensed in Mexico, Japan, and India.

More recently, Cummins has established joint ventures in India and Mexico and licensed production in South Korea and China because these and other developing countries could not afford large-scale imports due to shortages in foreign exchange. In parallel, the company also decided to develop a new line of engines of lower horsepower range (50 to 250 horsepower, B- and C-series) than the traditional Cummins products which are in the 250 plus horsepower range (NH-, V-, and K-series engines). The B- and C-series engines were about to go into production after several years of development work in a joint venture with the J. I. Case Company.

The world market for the higher power engines was estimated to be about 500,000 and that for the lower power engines was believed to be several million units a year. In the view of the company's leadership, it was impossible to supply such a large market from U.S.-based factories alone. Cummins' desire to locate foreign partners interested in co-production dovetailed well with the aims of developing countries to acquire technology from major manufacturers of industrialized countries by effective transfer mechanisms. Specifically, it was judged important for developing countries to absorb diesel engine technology, since this technology was essential for the development of their infrastructures. Further, Cummins was conditioned to taking a long-term view by the nature of its product: it has usually taken 5 years to design a new diesel engine and an additional period of 10 to 20 years to develop a market. This long-term planning approach was considered appropriate for operations in developing countries, where conditions and specialized resources were, in general, not readily available and first had to be mobilized or developed over a period of several years.

Cummins was prepared to transfer technology as an integral part of cooperative ventures but concluded that the technology transfer process of and by itself was not sufficient to make the venture profitable. The price that developing countries were, in general, willing to pay for existing technology was barely sufficient to cover the appreciable costs of the technology transfer process, in the view of Cummins executives. To make a profit, Cummins needed either to sell large quantities of parts or to share in the profits of sales in the country where its technology was manufactured.

DEVELOPMENT

China in Cummins' World Strategy

The strategy adopted by Cummins in developing countries was conservative and sometimes involved a defensive posture. The company was willing to take only limited risks. This was the strategy pursued in the Cummins China operation. In the company executives' view, there were no "windfall profits" to be made in China such as building up a manufacturing and marketing concern and realizing it later at large profit. Rather, income was expected to be on the interest return level; in other words, profits were expected to be modest and a long-term view was considered appropriate. Nevertheless, Cummins' world strategy called for penetration of the China market because of its considerable potential in terms of size of the country and because the Chinese government was considered to be basically prudent and honorable in financial matters. However, the company did not yet know how it could best participate in China's socialist economy.

China's industrial market was seen as having high potential.[4] Portable power was considered to be basic to China's growth, and the time was ripe for the development of road transport in addition to the off-road trucking needs. In 1985 most on-road trucks were gasoline powered, and it was realized by the authorities that the economics of trucking required conversion to diesel (dieselizing) of the existing fleet. Consequently, a sizable market already existed in the conversion process, quite apart from the expected equipment expansion.

China was not the only country in which Cummins was interested in placing its technology. Argentina, for instance, was also considered attractive in terms of its size and because of its large agricultural economy. However, Argentina was considered appreciably more risky, partly because the venture under consideration would require an equity infusion of $15 million. The final decision on this project was still pending at the time of the case study team's visit. Hungary was another country that had expressed interest, but in this case, Cummins decided not to invest in a government-subsidized venture amid a centrally

---

[4]The potential of the Chinese consumer market, by contrast, was judged as extremely low because of the low per capita income.

controlled economy of limited size. By comparison, China was considered appreciably more attractive, and in the judgment of Cummins' executives, the long-range outlook for profits there was good.

From the beginning, Cummins considered licensing only in China. The strong central government control over the economy in general and over the allocation of materials deterred the U.S. company from making an equity investment. In addition, the duration of equity joint ventures with foreign companies was in 1978 limited to 20 years. Although even this defensive posture was considered risky at the time of the inception of negotiations in 1979, Cummins sought a means of entry into the China market without incurring great cost and without exposing the company to excessive risk. It was thought that the China market might not reach economically advantageous dimensions for 15 years, but the country had decided to develop its diesel technology and was expected to do so with or without Cummins' participation. It was, meanwhile, the aim to establish Cummins in China as a reliable and trustworthy supplier of technology. To do this effectively, it was judged, the company had to begin laying the foundations for its position in China, since more aggressive competitors would otherwise gain a decisive advantage. Meanwhile, the Cummins licensing agreement with China was designed to promote the sale of a considerable volume of parts and, as a result of the cooperation, to promote the direct sale of finished products of the U.S. company as well. The latter expectation has been fulfilled to a considerable degree.

Historical Background and Early Contacts

During the early 1960s, a Japanese manufacturer, Komatsu, sold bulldozers to China that were powered with Cummins engines built under license (Marlow, 1985). Presumably, as a result of experience with these products, together with China's recognition of the company's world leadership position in diesel technology, a top company delegation was invited to visit China in 1975 (Marlow, 1985). J. J. Miller, the then chairman of the Board, led this group. During the visit Miller acquired a diesel engine produced at the Shanghai Diesel Engine Works. Cummins' personnel thoroughly studied and evaluated this engine to determine the level of Chinese diesel technology, which was, at the time, 20 to 30 years behind world standards. Richard Stoner, vice chairman of the Board, who later participated in the China negotiations, also was a member of the 1975 delegation. Following this visit, a number of mining trucks powered by Cummins' K-series engines were sold to China by a U.S. manufacturer. Cummins provided service training and spare parts for this transaction. However, the company did not invest any further staff time on research or planning concerning China during this interim period.

In early 1978, the China National Machinery Import and Export Corporation (Machimpex), which is subordinate to the Chinese Ministry of Foreign Economic Relations and Trade (MOFERT), purchased ten K-series engines directly from the Cummins Manila office. These engines were destined for a factory in Benxi, Liaoning Province, to power 60- and 85-ton off-highway trucks. Perhaps, as a sequel to this transaction, Cummins received a telex in July 1978 from the China National Technical Import Corporation (Techimport), also a MOFERT agency, charged with the importation of technology. Techimport invited a delegation to visit China in order to discuss the possibility of establishing a technology transfer affiliation. Probably, China's interest in Cummins was also strengthened by the latter's successful history of transferring technology to Mexico, India, and Japan.

Cummins responded to the Techimport invitation with enthusiasm; under the leadership of John T. Hackett, executive vice president and chief financial officer, and Richard B. Stoner, vice chairman of the Board, a group including ten technical specialists visited China in August 1978. Andrew Chu, an experienced Cummins engineer who had had his undergraduate education in Taiwan and received his Ph.D in mechanical engineering from Wisconsin University, was a member of the delegation. Chu played an important part as language monitor and occasional interpreter during the discussions. The Chinese hosts and counterparts during the two-week visit were the Automotive Bureau and the Mining Equipment Bureau, both of the Ministry of Machine Building Industry (at the time called the First Ministry of Machine Building), and Techimport (Marlow, 1985). The latter was represented by its Third Department, whose areas of responsibility comprised electric power, gas turbines, and transportation. Altogether, about 30 Chinese representatives participated in the meetings with the visiting Cummins group.

The purpose of the August 1978 meeting in Beijing was to explore possibilities of the transfer of Cummins' diesel engine technology to China and to develop an agenda for future contacts and activities. Cummins' specialists gave seminars and presentations to introduce and explain their products and answered questions posed by the host delegation members. It was agreed that the next step in the Cummins-China interaction would be a visit to China by a team of manufacturing specialists for a tour of those plants selected by the Chinese as potential recipients of the technology. This visit took place during September 1978.

Chinese Industry and Markets

In 1978, the independent information available to Cummins on the Chinese industry and China markets was minimal. The first delegation in August 1978 was briefed by their hosts, and the visitors accepted the market information so received as a basis for their evaluations and feasibility estimates. Cummins was told that the Chinese market for the diesel engines to be produced in China under Cummins' license was estimated at 26,000 units per year. During the ensuing negotiations, it became clear that the initial estimates of demand were unrealistically high; the numbers soon shrank to 6,000 and finally to even lower levels when the amounts of licensing fees were finally discussed. Even in 1985, with much more information available than there was in 1978, Cummins' personnel believed that the market demand was still poorly known. Moreover, the Cummins negotiators, during their numerous visits to China, were quite frustrated in that they failed to obtain any further significant information concerning the Chinese market or industry. One participant aptly expressed this frustration: "It felt like there was a vacuum on the other side which sucked up all information from the Cummins side without giving anything in return." Some information concerning the Chinese diesel engine industry was available to Cummins early on in the wake of the 1975 visit. More details on a few selected plants were obtained by the team of four manufacturing experts, headed by Jerome Schlensker, who visited China in September 1978 as a direct result of the exploratory visit by the Cummins delegation the preceding month. It was the aim of the September 1975 visit to gather information for making feasibility and cost estimates. The team was met by a five-member Chinese delegation, including an interpreter. Accompanied by this group, the Cummins team visited the Chongqing Automotive Engine Plant located in Chongqing, Sichuan Province, in China's central heartland; a number of supplier plants located nearby; two other diesel engine plants; and a "green field" site at Weifang, a small town in Shandong Province on the East coast (Marlow, 1985). A new plant was to be built at the latter site, according to the original Chinese plans, to produce the Cummins' NH-series engines; the Cummins' K-series engines were to be produced at the Chongqing plant. The latter factory, it was learned, had developed a tentative "ambitious plan calling for the construction of a new technical center, new assembly and test building, a new building for turbocharger production, as well as expansion of existing buildings" (Marlow, 1985, p. 12). Altogether, the total floor area of the plant was to be expanded from 50,000 to over 100,000 square meters in preparation for the introduction of the new Cumins technology. The management of the Chongqing Automotive Engine Plant requested input from the visitors to help finalizing these plans in anticipation of the affiliation.

The major concerns reported by the Cummins team regarded the level of quality control and a general lack of cleanliness at the plants visited. They also confirmed the conclusion made by the 1975 delegation: The Chinese diesel engine industry was 20 to 30 years out of date. The Chongqing factory was producing a 1960's vintage French Berliez engine, the technology for which has been acquired from France in 1965 under a "static" license agreement, that is, a one-time supply of paper containing design and manufacturing information without any further update or other contact.[5] Production of a larger 12-cylinder turbocharged engine with a horsepower rating of approximately 700 was in its in-

---

[5] The Berliet engine is a 6-cylinder, 180- to 250-horsepower unit, similar in physical dimension to the Cummins' NH-series engine but lower in power output.

itial stages, and it was understood that this product was to be replaced by the Cummins K-series engine (Marlow, 1985). The visiting team evaluated the plant's facilities, determined the changes deemed necessary for the production of the K-series engine, and submitted suggestions for improvements to the Chinese hosts in regard to industrial waste treatment, office space enlargement, and noise-reduction engineering.

Political Environment and Its Changes

In 1978, at the beginning of serious negotiations with China, a political risk analysis showed that the risk was not too great for further progress in the straight licensing venture, since the Chinese government was viewed as honorable and following prudent financial policies. This, together with the potentially large internal market, made entry into the Chinese economic scene attractive enough to outweigh the qualms concerning possible future political instability.

It should also be noted that China had already assigned high priority to the development of natural and energy resources. The central government would thus be certain to encourage the development of a sizable market for high horsepower Cummins engines, since these were required to power the heavy off-highway vehicles used in exploration and mining activities. In addition, Cummins' executives believed that the transportation sector occupied a place of prominence in the governmental development policy and could in time offer extensive opportunities for the new low-horsepower (below 250 horsepower) Cummins line to power on-highway trucking.

Delegations and personnel. The personnel involved in the Cummins China venture were led, from the very beginning, by two executives of the highest corporate level: John T. Hackett, executive vice president since 1978 and chief financial officer, and Richard B. Stoner, vice chairman of the Board. The involvement and support of these two senior executives were considered crucial. Their direct participation in many of the interactions with the Chinese counterparts resulted in their understanding of and appreciation for the difficult problems encountered by the negotiators. In addition, Stoner had been a member of the 1975 delegation, and he and Hackett had participated in the first exploratory discussions in Beijing in August 1978.

When initiating a new affiliation, Cummins distinguished between the development team, the negotiating team, and the start-up team, although there was considerable overlap between these groups in the present case. The personnel for all teams were drawn from the company's functional groups, viz. engineering, manufacturing, and business, and were chosen for their status of professional credibility within their respective groups. The individual must be able to accurately represent the company in China and then to go back to home base to draw on its support as required. Andrew Chu, general manager of the China Business Group, described it as follows:

We draw upon a very strong person in each functional area so that he can speak for the company in front of the customer and so that he also has the credibility (with his group) and the connections to go back to his own organization (to draw on its support) and so that all commitments (made by him) are felt to be the company's rather than the individual's commitments.

NEGOTIATIONS

Venue and Personnel

All negotiations between Cummins and Techimport took place in Beijing. The timetable for negotiation meetings and other key events of the Cummins-China interaction process is presented in Exhibit 1 on page 166. MMBI's personnel and representatives of the plant that was designated for production of the diesel engines were present at the negotiations but did not actively participate. Presumably they served as silent partners to the Techimport negotiating team and kept their home units informed concerning progress in the negotiations. The designated production site for the diesel engines was the Chongqing Automotive Engine Plant located in Chongqing, Sichuan Province, the largest city in China. This factory was subordinate to the MMBI's Automotive Bureau, now called China National Automotive Industry Corporation (CNAIC).

Negotiating team personnel were selected by Cummins using the following criteria: (1) in accordance with general company policy described above; (2) having the ability to reach across cultural lines and close the intercultural gap; (3) having the ability to understand the thoughts expressed by a person who speaks another language and who lives and works in an entirely different cultural setting. The Cummins negotiating team included two full-time members: Keith Chambers and Dennis Kelley, who both had experience in handling foreign assignments, the former in the Far East and the latter in the Middle East. The third and part-time member of the team was Andrew Chu, whose primary responsibility between January 1979 and January 1981 was as director of Engineering for Asia-Pacific, succeeding Keith Chambers. He worked out of the Cummins Singapore office. Chu was called in to join the negotiating team in China whenever possible. Chu was the only member of the negotiating team who understood both the Chinese language and the culture, although he did not, originally, have first-hand knowledge of the sociopolitical environment of the People's Republic.

Chu believed that the ability to listen was paramount and far more important than familiarity with the particular culture of the counterparts in negotiation. Nevertheless, Chu was credited with making a very significant contribution to the negotiation process, as illustrated by the following example, reported by Marlow (1985). At one point when Chu was not present, the negotiating team encountered extreme difficulties arising from misunderstandings due to poor translation. When Chu was called in, matters were soon resolved, and the negotiations once again were able to progress on an even keel.

Chu described the nature of these hurdles to effective communications that were related to cultural differences. For example, the unkempt appearance of the Chinese negotiators, seemingly plain or even sloppy clothing, unshaven faces, and so forth, was taken as an indication of low level capability by the U.S. negotiators. In another case, the term *strategy* was used by a Chinese negotiator, causing the U.S. side (presumably Chambers and Kelley) to jump to conclusions concerning far-reaching strategic decisions made by the Chinese side. However, the remark was most probably much more casual and lacked the deep meaning ascribed to it. Chu used these examples to emphasize how important it was to listen to the Chinese negotiators in order to understand their habits, mannerisms, and the significance of terms as they use them. He stressed that the ability to listen is a very rare commodity in U.S. corporations, where most executives and managers tend to assume that it is the job of others to listen to them. Chu summarized his own contribution in the following terms: "My major contribution was to understand the words and meaning and to translate them into the Cummins context."

Negotiating Styles

During the initial phase of the contacts between the sides, beginning with the Cummins delegation's first visit to China in August 1978, much time was spent on information exchange. Cummins' personnel described the company's products and its policies in the areas of product development, sales, servicing, and international trade. Cummins' technical personnel were responsible for convincing the potential customers, that is, the Chinese, first and foremost, of the high quality and desirability of the company's product and of its underlying technology. On the other hand, the Chinese engineers described their views of needs and requirements. These discussions and exchanges between technical personnel of both sides were successful, according to Cummins' participants. They said that the Chinese engineers were on a par with their Cummins counterparts as far as their level of technical expertise was concerned and that communication between the two groups was satisfactory.

Beginning with the second meeting in December 1978, a number of concepts were introduced by the U.S. side that had to be explained to the Chinese negotiators. It was quickly agreed between the parties that the projected technology transfer was to be conducted under a licensing agreement without any equity investment by Cummins, since neither side was prepared to consider any other alternative. However, Cummins introduced the concept of a "dynamic licensing agreement," which was new to the Chinese negotiators and therefore required a great deal of explanation (Marlow, 1985).[6] As a matter of policy, Cummins only entered into dynamic licensing agreements, since static agreements were not considered appropriate when the licensor wished to enter into the long-term relationship with the licensee and wished to establish itself as a solid and reliable supplier of technology. The commercial terms of the agreement generally required a great deal of discussion, since the Chinese counterparts in the negotiations were not well informed on internationally accepted commercial standards. These difficulties were encountered because the negotiations were carried out during an early period when Techimport personnel had little experience with licensing agreements.

At one point, Cummins negotiators were told indirectly by Automotive Bureau personnel who attended the negotiations that the Chinese side was not negotiating with any other party for diesel engine technology transfer. The Chinese thereby gave away a trump card, but Cummins decided not to make use of this information to gain any advantage. By this time Cummins was reasonably sure that no other company was willing to accommodate the Chinese side to a similar extent as it had. It was known that Caterpillar Tractor Co. had pulled out at an early stage of discussions when it became clear that they could not establish an engine plant in China and hold controlling interest; Caterpillar was not prepared to settle for a licensing agreement. Also, Mercedes ran into problems in its discussions and pulled out. According to Cummins personnel, the other companies "couldn't get the momentum going," probably because they didn't have the advantage of the long-range vision and direct involvement of the corporate leadership, which was so essential in the Cummins case.

Cummins enjoyed another special advantage over competitors, since its vice president of Research, Dr. W. T. Lyn, had high-level contacts in the Chinese bureaucracy and among the Chinese technical elite. Lyn had previously been professor of internal combustion engines at Imperial College, London, but he had begun his education in China and was a graduate of the prestigious Qinghua University in Beijing. As a result, he was well acquainted with the Chinese engineering community and its leaders; his name served as a reference for Cummins and helped to establish credibility for the company within the Chinese power structure. In 1980, Lyn succeeded in arranging a visit by the minister of the Machine Building Industry to Cummins' corporate headquarters. Since the visit occurred during the period of negotiations with China, it was of special symbolic significance since it sent a visible, favorable message to the Chinese bureaucracy.

---

[6]Under a dynamic licensing agreement the licensor undertakes to keep the licensee informed concerning changes and updates on the technology which is being licensed, as opposed to the situation in a static licensing agreement which involves a one-time transfer of technology.

Cummins personnel said that they perceived the Chinese negotiators as "hired hands without vested interest." The Techimport negotiating staff seemed to them to be divorced from the actual substance of the negotiations, resulting in needless tensions and lack of directed, meaningful purpose. As a result, the negotiations were described as generally unpleasant by Cummins participants. They reported that there was a persistent feeling that they were "shadow boxing" with the Techimport negotiators. There often was a feeling of "softness" and "mushiness" or of uncertainty concerning the subject matter being negotiated. Cummins negotiators ascribed this general malaise to the fact that their Techimport counterparts were not connected with the manufacturing group and were not well informed concerning the issues relevant to the quality and level of the technology under discussion. As a result, they were not able to make the connection between the words used and the content in terms of manufacturing requirements and the means of technology transfer to be employed.

It has already been noted that representatives of the recipient factory and the relevant ministry were present at the negotiations but were not directly involved in them. Altogether 40 individuals were involved in the negotiations on the Chinese side during the entire two-year period. The imbalance in numbers of negotiators on the two sides is remarkable.

In the view of the Cummins negotiators, the Cummins team handled the negotiations well; Chambers and Kelley did most of the negotiating in Beijing with the support of Chu. Only at the stage of settling financial matters, specifically prices, was Hackett brought in and at this stage Chu took charge of translating. The Techimport personnel seemed to be continually surprised by the speed with which Kelley and Chambers were able to make decisions. The enlightened leadership of the Cummins management and in particular Hackett's active interest in the process are credited with creating favorable conditions by giving the negotiators considerable authority.

Additional Support

When no Cummins personnel were in China, additional support was obtained from Sam Wang of the International Corporation of America, whose headquarters is in Washington, D.C. Wang has lived in Beijing since the early 1970s, and he had served irregularly as an informal contact for Cummins since the mid-1970s. In 1978, a formal contractual agreement was entered into, and since then Wang served as resident Beijing contact until the establishment of Cummins' Beijing office in 1984. Sam Wang also sat in on the meetings between Cummins and Techimport. Otherwise, apart from minor consultations, the Cummins team had no outside support during the two years of negotiations.

Pace of Negotiations

According to information obtained from Cummins personnel involved in the process, the pace of the negotiations was entirely controlled by the Chinese from the onset of negotiations in August 1978 to the crucial meeting in August 1980. Long hiatuses occurred following some of the negotiating sessions, and at one time in 1979 several months went by without any communications between the sides (Marlow, 1985). In fact, Cummins negotiators Kelley and Chambers at that time concluded that the Chinese side had decided to go elsewhere and had broken off the negotiations when they were not successful in obtaining any response from Techimport, even through their resident representative, Sam Wang. They even requested reassignment at one point, but John Hackett prevailed on them to stay with the project until more concrete evidence of suspension of the negotiations could be obtained (Marlow, 1985).

Financial Considerations

Hackett, accompanied by Stoner, negotiated the cost of the technology and of components to be sold by Cummins. The first time a pricing proposal was presented in December 1979, the Chinese team, after a brief huddle, "put on their coats and left the room," according to Marlow (1985, p. 19), ending the four-day meeting. This move was interpreted by the U.S. side as signifying extreme displeasure with the proposed terms. In this instance, however, communications between the parties were quickly reestablished in January 1980 at the Chinese side's initiative. At the next meeting in March 1980, a new pricing proposal was presented by Cummins involving a complex formula for royalty payments, but no final decision was made (Marlow, 1985).

Meetings between the parties were not resumed until the following August. At that session the Techimport team responded to the Cummins proposal originally presented in March by stating that the cost was too high, but they refused to come forward with a concrete counter proposal. On that occasion they also were reported to have engaged in "unprofessional behavior" by questioning Cummins' payroll policy and the validity of Cummins' salaries, which were criticized as being exceptionally high and as unduly increasing costs (Marlow, 1985). John Hackett and Richard Stoner were present at that session, and as high-level corporate executives, they were in a position to break off the negotiations. Thus the Cummins team summarily left China, even though such a move was considered risky at the time by at least some members of the team. Nonetheless, it made a deep impression on the Chinese negotiators, who were quite surprised by Cummins' move. That this decision was made by senior corporate officials was not lost on the Chinese. This action by Cummins marked a turning point in the negotiations since after that Techimport's approach to the negotiations appeared to be significantly more serious. The U.S. company personnel also felt that after this incident they controlled the pace of the negotiations.

In addition, Cummins speculated that Techimport was coming under pressure from the MMBI to expedite the negotiations and to arrive at an agreement. Financial considerations were not discussed again until January 1981.

This time the atmosphere was much improved. Cummins set a time limit for the final negotiations and stuck to it in the face of strong pressure on the part of the Chinese to extend their stay. Finally, the contract was signed despite Chinese reservations concerning some of the terms. The occasion was strained and lacking in jubilation (Marlow, 1985).

An outside but highly relevant circumstance served to expedite the completion of the negotiations. In 1979, the Chinese concluded an agreement with the Japanese firm Komatsu under which two factories in China were to produce bulldozers under license. Komatsu's own technology was limited to the manufacture of bulldozers, but they powered their product with Cummins' diesel engines, which Komatsu in turn produced under license. In fact, this was one channel through which the MMBI had become acquainted with Cummins and its technology as well as its reputation as a successful technology transfer agent. Now, Komatsu's licensing agreement with Cummins did not permit the former to sublicense Cummins' technology. As a result, the Chinese were highly motivated to conclude an agreement with Cummins, since only the Cummins engine was fully compatible with the Komatsu equipment and the bulldozers were about ready to roll off the Chinese assembly lines.

Once Cummins became aware of this situation, it was naturally less worried about losing out to competitors. This realization contributed also to its willingness to incur greater risks, and consequently its negotiators came to adopt the firmer negotiating style described above.

Choice of Production Plant

Cummins' information concerning Chinese diesel plants was derived from an inspection trip of a Cummins team under the leadership of Jerome Schlensker in September 1978. This team visited several plants, critiqued what they found, and reported that the Shanghai Diesel Works, which is subordinate to the Ministry of Agriculture, was the best of these. However, when it came to negotiations, Cummins never had input in the matter of choice of a recipient plant for its technology. For one, the negotiating partner was the Ministry of Machine Building Industry, and there was then and still is now no way of crossing the administrative boundaries between Chinese ministries. Nor was Cummins allowed any input in choosing among the plants operated by the MMBI.

Cummins personnel saw the process of selection of the recipient plant as an internal process within the "monolithic Chinese structure," in which different plants competed under rules still not understood by them. Cummins was presented with the Chinese decision to manufacture the engines under license at the Chongqing Engine Plant. The U.S. side then decided in response that this was an acceptable choice. According to Cummins participants in the negotiations, their company would have preferred to have input on the question of choice of plant; for example, they would have preferred a more accessible site such as Shanghai.

Originally it had been the plan of the MMBI to build a new factory in Shandong Province under the administration of the Ministry's Mining Equipment Bureau to produce the NH-series engines. However, Cummins was informed in July 1980 of the Chinese decision to produce both the NH- and K-series engines at the Chongqing plant, which was subordinate to MMBI's Automotive Bureau. This latter decision was ascribed by Cummins to the economic "readjustment" policy, in its initial stages of implementation at that time. Basically, the Chinese government was implementing a policy of general retrenchment, and new construction projects in particular were to be kept to a minimum. Cummins personnel also speculated that some internal politics between the Automotive Bureau and the Mining Equipment Bureau may have played a part in this decision-making process (Marlow, 1985). In this case, Cummins was consulted and was asked to send a team to evaluate the feasibility of producing both engine series at the Chongqing plant. Schlensker and Kelley made the trip to Chongqing in late July 1980 and reported that the new plan could be carried out, although they found that some difficulties would have to be addressed (Marlow, 1985).

Conclusion of Agreements

The licensing agreement was signed in January 1981 between Cummins and Techimport. As already discussed, the circumstances were such that nobody felt jubilant after the long, drawn-out process and particularly since the very last negotiating session was marked by tension and confrontation. However, the relations between Cummins and the Chongqing Plant, the actual licensee and the factory where the engines were to be produced, were not affected by this tension, since the negotiating personnel on the Chinese side seemed to be quite divorced from the plant's personnel.

Cummins personnel attached special importance to the licensing agreement they negotiated, since, according to them, it was the first agreement of its kind concluded between a foreign company and China. This agreement subsequently served as a model for later contracts. Cummins did not provide the authors with a copy of the agreement, and only selected particulars were discussed during the interviews with the company's management personnel. However, an outline of the contents of the agreement is described in Marlow (1985). According to Marlow, Appendices 6 and 7 of the agreement are the most important. The former contain lists and definitions of all patented technology with dates of expiration of the Cummins patents. This was important since the Chongqing Plant was licensed to use the technology inherent in the NH- and K-series engines for the period of the license agreement only. The agreement, as signed in 1981, was valid for ten years and was renewable. However, if a patent was to expire before the termination of the license agreement, the licensee would obviously be free to continue exploiting the technology contained in this patent. Appendix 7 specified technology contained in the relevant Cummins engines that was produced by other companies and for which other companies owned patent rights. This part of the agreement had proved to be the most commonly referred to during the Cummins–Chongqing Plant relationship, since the Chinese side had been reluctant to accept the fact that Cummins was unable to provide details concerning such technology. Cummins adopted the strategy to facilitate the Chongqing Plant's contacts with the original suppliers of the parts concerned, so that the Chinese could negotiate directly with them concerning supply or production (Marlow, 1985).

According to Cummins personnel, the agreement specified that the quality of the Chinese product was to be similar to that of the original Cummins engine built in the U.S. plant and subject to the same quality inspection guidelines. There were no provisions in the agreement for the enforcement of quality standards, however, and Cummins had no legal means to enforce its quality control procedures. Cummins personnel pointed out that the technology was licensed in exchange for cash payment and not for product to be supplied to Cummins. If the latter were the case, Cummins would have had control, since the product would then have had to meet company quality standards.

According to the agreement, all parts to be supplied to the Chongqing Plant had to be shipped by Chinese flag carriers. This limitation resulted in many delays and caused some difficulties with meeting schedules. The agreement constrained China from exporting engines produced under the Cummins license, but there was no restriction on indirect exports, that is, on exports of equipment containing the engines as built-in components such as tractors, trucks, and so forth. The freedom regarding indirect export posed some threat to Cummins, inasmuch as lower quality products could be injurious to Cummins' reputation in the international market. The agreement also specified criteria for the quality of material and components sourced in China by the licensee; the locally sourced parts were to meet the "Cummins source approval test" requirements. This condition was later considered to be too severe by the licensee and was renegotiated to the Chinese side's satisfaction, according to Cummins personnel.

During and after the negotiations, both sides were subject to some internal pressures, according to the Cummins personnel interviewed. At Cummins there was resistance on the part of the engineering staff to the sale of the technology developed by them. They felt that the technology had been sold at too low a price. This stance was opposed by the commercial staff, who were fully aware of the intensity of competition in the international market place. Internal tensions were successfully eased by holding meetings to fully explain the advantages and risks of the licensing agreement with China. On the other side, Techimport came under intense criticism after the signing of the contract in January 1981 and was accused of having concluded an agreement that was disadvantageous to China. As a result, pressure built up on the Chinese side to press Cummins for renegotiation of the financial terms. Such renegotiation did, in fact, occur during the start-up phase.

START-UP

Personnel and Organization

At Cummins the negotiation of cooperative venture agreements, including license agreements, was usually assigned to a small group or team including technical, commercial, and legal experts. The same procedure was followed in the case of the agreement with China's Techimport except that, in deference to Chinese sensitivities, legal expertise was not represented on the official Cummins negotiating team. It was further usual to turn over the task of starting up the new venture to one of the existing organizations within Cummins, the choice depending on the type of agreement and on the type and location of the counterpart or licensee. This procedure was not followed in the present case, since it was felt that the cultural and technological differences between Cummins and the Chongqing Automotive Engine Plant were greater than had been encountered in previous company licensing experience. It was therefore decided to provide for continuity of personnel from the negotiating process and to form a special unit called the China Business Group which was established in February 1981. Andrew Chu, the experienced Cummins engineer of many years who had also provided the language and cultural link with the Chinese negotiators during the long process of negotiation, was selected to head the China Business Group as its general manager. Dennis Kelley, the full-time member of the negotiating team since its inception, was named director of operations of the China Business Group.

The tasks assigned to the China Business Group included, as a matter of course, the facilitation and administration of the license agreement implementation, including all aspects of these functions. In addition, the unit was also charged with developing the China market, initially for the products of the Cummins-licensed Chongqing Automotive Engine Plant. In 1985 the China Business Group was also charged with the entire marketing and direct sales effort of Cummins in all of China, including developing and negotiating new business ventures. It was originally planned to phase out the China Business Group after the third year of its existence, by which time its functions were to be absorbed into the existing Cummins organizational structure. However, in 1985 the China Business Group was still fully engaged in developing additional joint enterprises with Chinese entities as well as administering the full scope of current undertakings, including the Technical Service Center, established in Beijing in 1984. The organizational chart presented in Exhibit 3 on page 168 also confirms the wider responsibilities assigned to the China Business Group.

As shown in the chart, the China Business Group had three divisions—Manufacturing, Engineering Liaison, and Operations. Originally, the China Business Group had 19 members, but in 1982 personnel cuts reduced the staff to 14, the number of manufacturing project leaders being reduced first. The project director position was filled by Roger Lang and, after staff reductions took effect, he was responsible for all relevant areas of expertise. Dennis Piper, a design engineer at Cummins for about 15 years, was named manager of Engineering Liaison. For four years preceding this appointment, Piper had worked in the International Business Management Group and had been an "implementer of international ventures" in India, Mexico, Korea, Brazil, and most recently Indonesia. He had therefore been engaged in "technical support work for plants around the world" for a number of years, but he had no China background. The third division of the China Business Group, Operations, was headed by Dennis Kelley and supported by Stephen Mulder, who was manager of Marketing.

It was the responsibility of Lang and Piper to prepare for the Joint Planning Session with their Chinese counterparts from the Chongqing Automotive Engine Plant, scheduled for the Fall of 1981. Piper was responsible for the transfer of technical data and design specifications for the NH-series engines, and Lang for the transfer of manufacturing plans in advance of the Joint Planning Session in order to allow the Chinese engineers to begin making plans for the manufacturing process in China. These two senior China Business Group managers were also responsible for the planning and subsequent execution of the training programs to be conducted at Cummins and for follow-up training in China. During the manufacturing start-up at the Chongqing Automotive Engine Plant, Lang and Piper were also to continue to be the resource persons through whom the licensee could obtain additional information and clarifications. They would also decide what additional documentation was to be sent as part of the dynamic licensing agreement.

On the other hand, Kelley's responsibilities as head of Operations were wider and extended beyond the immediate licensing project. In fact, all marketing and business activities of Cummins in the People's Republic of China were transferred to the China Business Group when it was established. As a result, China Business Group's Operations division became responsible for direct marketing in China as well as applications engineering and servicing for direct sales of Cummins' products. This was in addition to its responsibility for the administration and administrative support for all activities connected with the licensing agreement with the Chongqing Automotive Engine Plant. Francis Chu, a native of China, who was educated in Taiwan, was hired to give administrative support to the technology transfer process.

Technology Transfer

Dissemination of documents. According to the agreement between Cummins and Techimport, the technical documentation supporting engineering design and manufacturing of the NH- and K-series engines, was to be sent to China (to Techimport for delivery to the Chongqing Automotive Engine Plant) within three months after obtaining the necessary export license from the U.S. Department of Commerce. This license was obtained in March 1981, a relatively short time after the signing of the licensing agreement in January of that year. The task of assembling and dispatching the relevant documentation proved to be a mammoth undertaking, which had been expected by the Cummins personnel. Negotiators had only agreed to the short time frame under strenuous pressure from the Chinese side.[7] Friction developed since Cummins found it difficult to send the large volume of paper by the formal time limit and also since the Chinese side demanded more documentation than Cummins personnel judged necessary for the licensed production. The problem may have been exacerbated by Techimport's demand for more material and faster delivery while being divorced from the actual process of absorption and integration of the information.

The particular aspect of the Cummins–Chongqing Automotive Engine Plant venture that caused an additional problem was the dynamic character of the licensing agreement. Neither the Chinese negotiating team nor the technical Chongqing Automotive Engine Plant personnel had had previous experience with this type of venture and therefore did not make allowance for the magnitude of the resulting quasi-continuous flow of documentation. Because of the Chinese focus on obtaining as much paper as possible from Cummins, the choice of documentation to be sent on a continuing basis became an issue of dispute. Quite apart from the divisive nature of the differences over the documentation issue, the dynamic nature of the licensing agreement presented ongoing administrative as well as technical problems for the Chongqing Automotive Engine Plant in the view of Cummins personnel.

Joint Planning Session. The Engineering Liaison and the Manufacturing divisions of the China Business Group were to develop training programs for Chongqing Automotive Engine Plant engineers; the Operations division was to design parallel programs for Chinese administrative and commercial staff. These plans were to be finalized in consultation with representatives from the Chongqing Automotive Engine Plant at the Joint Planning Session, which was held in Columbus in the Fall of 1981. In preparation for these meetings, Roger Lang, project director of the Manufacturing division, and Dennis Piper, manager of the Engineering Liaison, traveled to Chongqing to visit the licensee in June 1981, almost immediately after the technical documentation had begun to flow from Cummins to China. The purpose of

---

[7]The Chinese are in general anxious to take possession of all technical documentation as quickly as possible, often out of proportion to their ability to assimilate that information. It is assumed that this wish is motivated by their past unfortunate technology transfer experiences. During the period 1950–1960 China was strongly dependent on the USSR for developing its industrial and technological base, and the Russians, although generally supportive, showed reluctance to turning over the full documentation to the Chinese trainees. When the split between China and the USSR occurred in 1960, the Russian experts and advisors left abruptly and took with them the documentation, which they had in many cases kept under their control.

the trip was to determine the compatibility of their tentative plans for the Joint Planning Session with Chongqing Automotive Engine Plant realities and Chinese objectives. At the same time, the visit served to establish personal relations between Lang and Piper on one hand and their counterparts on the other and to probe Chinese understanding of the relevant Cummins products. The Americans were impressed with the intensity with which the Chongqing Automotive Engine Plant personnel had studied all available published information. They had also located and borrowed Cummins engines from a Chinese source to study their structure and function. Based on these studies and on studies of the newly received documentation, the Chongqing Automotive Engine Plant engineers had developed preliminary manufacturing plans. During their visit, the Cummins engineers also introduced their counterparts to Cummins' terminology and indexing systems in order to facilitate future communication. At the same time, they acquainted themselves with the organization and procedures of the Chongqing Automotive Engine Plant.

The Chinese participants of the Joint Planning Session were divided into seven groups. Two groups were composed of bureaucrats and technocrats, the first representing Techimport and the Ministry of Machine Building Industry (referred to as the *executive group*) and the second (called the *administrative group*) was composed of top-level officials from the Chongqing Automotive Engine Plant. Four groups were composed of senior engineers of the Chongqing Automotive Engine Plant and dealt with most technical aspects of the project. The last group was composed of engineers of the Chongqing Fuel Systems Plant, which is a supplier of the Chongqing Automotive Engine Plant. In preparation for the Joint Planning Session, the Chongqing Automotive Engine Plant technical groups were to assess the Chongqing plant's current capabilities for the proposed manufacturing process and to prepare preliminary manufacturing plans as basis for discussions with Cummins manufacturing experts who had been recruited by the China Business Group.

The Joint Planning Session was to be the major kickoff event for the implementation of the licensing agreement, and all aspects of the project were to be decided upon during this series of meetings between the executives and managers of both parties. In addition, the Joint Planning Session was designed to be an opportunity for the Chinese officials and engineers to become acquainted with Cummins' facilities and methods in the United States as well as in the United Kingdom. Visits to plants in the U.K. were considered of importance in order to broaden the Chinese engineers' and managers' knowledge of Western industry. Also the licensee was to receive deliveries of parts and CKD (completely knocked down) units from these plants. In fact, the Chongqing Automotive Engine Plant technical groups began their tour in the U.K., where they visited plants of Cummins' affiliates. They then proceeded to the United States and toured plants of Cummins and of Cummins suppliers in South Carolina, New York, and Ohio before arriving in Columbus, Indiana, in September. The foundry group arrived in Columbus in time to join the others in a week's tour of the Cummins facilities and then in a trip to a number of foundries in the Unites States, returning to China via Japan to tour the Komatsu foundry.

The Joint Planning Session began with general meetings at which both sides gave presentations to acquaint the other partner with their respective organization. The technical Chongqing Automotive Engine Plant groups, consisting of four or five engineers each, convened with Cummins' manufacturing project leaders to develop manufacturing plans. These technical sessions extended over six weeks. It soon became apparent that the language barrier was considerably greater than Cummins' personnel had anticipated, and communication was largely limited to using interpreters. Nevertheless, both sides were enthused by the prospects of a successful venture and felt that the six weeks had been profitably spent.

The executive and administrative groups arrived in Columbus after touring the Charleston, South Carolina, Cummins plant to hold consultations with the Cummins leadership. The technical groups had by then completed their work, and one member of each of these groups remained to join the executive consultations.

The first item of business was a Chinese request for renegotiations of some of the terms of the agreement. After some initial trauma on the Cummins side, this problem was overcome, and eventually all outstanding matters were settled satisfactorily. According to Marlow (1985, p. 54):

> By the end of November . . . the two partners had ironed out many of the differences and eliminated many of the uncertainties. Perhaps more importantly, the Joint Planning Session, as Cummins had hoped, [had imparted] a more personal flavor to the partnership and laid the foundation for technical and commercial cooperation.

Renegotiations.   The first time Cummins encountered a request for renegotiation of terms was during the Joint Planning Session (Marlow, 1985). The management of the Chongqing Automotive Engine Plant had determined, after examining the agreement as it had been negotiated by the Techimport representatives, that it contained some conditions which they found impossible to meet. This was evidently the result of the lack of coordination between Techimport and the plant that was to implement the agreement. The Cummins personnel were shocked by this demand, since they assumed that it was the responsibility of the negotiators to ensure the viability of the terms of the agreement. In the end, it turned out that the changes requested were not fundamental and the episode ended with everybody satisfied, as already noted (Marlow, 1985). Nevertheless, the incident illustrated the differences in concept of a concluded agreement in the minds of the two parties. Cummins assumed that a signed agreement was sacrosanct, whereas the Chinese side obviously assumed that the agreement could be renegotiated if the original terms were unsuitable when it came to implementation.

Later, after start-up of production at the Chongqing plant, the Chinese licensee concluded that the agreed Cummins approval test for sourced products was too severe for China, and Cummins was requested to renegotiate also this issue. A less stringent new procedure was established which the U.S. side still considered adequate, according to Cummins interviewees. However, Cummins insisted that its personnel must be present during the testing procedure, and this condition was agreed to after some considerable resistance by the Chinese side.

More painful to Cummins personnel than the preceding incidents was pressure from the Chinese to renegotiate the financial terms of the licensing agreement in 1982, during the start-up phase. Cummins viewed the agreement as having been settled after two and one half years of often difficult negotiations, and the U.S. side resisted reopening the negotiations barely a year after their conclusion. The Chinese designated the terms of the original agreement as "unfriendly" and pressed for lower prices for Cummins components. Cummins had already lost revenue since the compensation to be received depended on the volume of engines produced, and the total had been reduced substantially below the original Chinese estimates. In the end, Cummins gave in and agreed to some reductions to save the venture, since the Chinese bureaucracy threatened to shut down the project.

Training Program

Cummins decided that it was inconvenient to conduct the training of Chongqing Automotive Engine Plant personnel at the corporate headquarters. Therefore, a special off-site facility was organized and equipped with the necessary blueprints, engine models, and other training aids. The training was entirely on the job, and no formal classroom instruction was included, except for some introductory sessions aimed at helping the visitors adjust to their new environment. These introductions were conducted by Francis Chu but were discontinued when the flow of trainees thinned out. Technical sponsors were designated for each group, and these sponsors were responsible for the visitors' individual problems. In response to inquiries from the study team, Piper explained:

> The trainees are responsible for asking questions and for adapting the acquired knowledge and skills to their environment, which they know. [The trainees] were never tested or checked as to how they planned to adapt [the learned] procedures to their environment.

However, the U.S. instructors did make an effort to understand where the trainees "were coming from" during question and answer sessions.

The trainees expressed some dissatisfaction with the style adopted for the training program. They expected more classroom and presentation type instruction, although the nature of the planned program had been clearly specified in the agreement. The wording used was that the training would be on-the-job type as used for new Cummins employees. However, this information contained in the agreement did not get passed on to the engineers sent for training. The Cummins personnel interviewed were in agreement that much of this problem arose from the difference in educational and training style in general use in the two countries.[8] In spite of these difficulties, the overall training program was viewed as having been successful. However, the number of trainees sent by the Chongqing Automotive Engine Plant to Cummins was considered to have been too small and, in fact, was appreciably smaller than foreseen in the agreement. Cummins personnel believed that the Chinese side cut back because of a shortage in foreign exchange, thereby also slowing down the technology absorption process.

The Chinese side had wanted to have the Chongqing Automotive Engine Plant research and design engineers trained first and proposed that the manufacturing personnel be trained only during the fourth year of the cooperation. It required considerable persuasion and negotiating to convince the Chinese side that manufacturing training should be first in order to enable the production process to begin at the Chongqing Automotive Engine Plant. Actually, design training began in 1985 and was limited to the presentation of the design rationale for those NH- and K-series engines scheduled for assembly and eventual production in China. This instruction was aimed at raising the capability of the Chongqing Automotive Engine Plant personnel to better understand the dynamic nature of the agreement between the parties and to absorb the ongoing design changes, which resulted in a continuous flow of paper from Cummins to the Chongqing Automotive Engine Plant. In this context, Thomas Head, vice president and general manager of Affiliated Enterprises, told the investigators that it was characteristic of Japanese and Chinese licensees and partners in other types of cooperative ventures to know more than just "what it is and how it is." He quoted them as asking,

> How did you figure out what this is, and why did you figure it out that way, and what are your means of designing, appraising, analyzing, and so forth?

Head continued:

> We never figure that this is really part of what we are selling. We go beyond documentation and hands-on training, but we feel uncomfortable about it, helping them learn all this other stuff.

It therefore appeared that Cummins personnel responded to deeper inquiries concerning design rationale with reluctance, since they felt that such responses went beyond their commitment.

---

[8]In China, education is largely based on the presentation of information which the student is to absorb, rather than the participatory style more usual in the United States.

The follow-up training on site in China was limited by the guarded attitude of the Chongqing Automotive Engine Plant. All Cummins personnel interviewed agreed that their Chinese partners wished to keep the U.S. personnel at a distance and wanted to do their own planning and development of the manufacturing process. The Chongqing personnel wanted to limit the interaction with Cummins engineers to the latter responding to specific questions posed by the Chinese. Even then, the U.S. visitor was usually consulted in a reception room separate from the facilities where the Chinese staff worked. This form of communication was considered limiting and inefficient.

In the view of Cummins managers, new personnel added to the Chongqing Automotive Engine Plant staff in the commercial area were not well trained and were narrowly focused. As a result, it was necessary for visiting Cummins personnel to train them during visits to the plant, a process costly in time and effort.

Overall, the Cummins technical personnel were impressed by the professional competence of the Chinese engineers but were somewhat disappointed with their proficiency in English and their general lack of ability to interact easily with U.S. instructors. The experience during the training program in this respect was similar to that during the Joint Planning Session. In some cases, the U.S. side had given input concerning the choice of personnel who should be sent to Cummins for training, but it was unclear if such suggestions were implemented. On the other hand, Chongqing Automotive Engine Plant managers have sought Cummins personnel input concerning the English language capability of candidates to be selected for training abroad by requesting them to interview such candidates.

Cummins personnel speculated that the Chongqing Automotive Engine Plant managers stress professional and educational qualifications in their selection of trainees, but they said that political qualifications were probably also considered. In some cases it seemed that an individual was selected more because "it was his turn and not because it was important for him to learn." It also appeared that on their return, some trainees were transferred to positions other than those for which they were trained by Cummins. There were even cases where the individual was retired while in training at Cummins! However, these cases were in the minority, and most of the returning trainees were believed to have been suitably placed.

Development of Manufacturing Process

According to Thomas Head, during a technology transfer process, the supplier of the technology is usually "in the driver's seat" as far as engineering is concerned, since the receiving company had chosen the supplier's technology. However, in the manufacturing process, the licensee should be in charge, since it knows its own conditions best, Head said. In general, it was Cummins' policy to allow variation in manufacturing equipment in different plants and subsidiaries, even where the company had full control, unlike other companies like General Motors. As a result, Cummins personnel held to the principle that the local factory staff should play the major role in developing the manufacturing process. Head believed, however, that the licensee usually needed a considerable amount of help and support from Cummins. On the other hand, he agreed with the opinions expressed by other Cummins personnel: The Chongqing Automotive Engine Plant was not eager to make use of Cummins' support. As he put it, "Chongqing is sort of on the far end of 'leave me alone, Mother, I'll do it myself.' " It was the generally held opinion at Cummins that the Chinese know best what works under their conditions. The Chongqing Automotive Engine Plant chose some local equipment and sometimes, according to the opinion of Cummins' staff, chose equipment at too sophisticated a level.

On the other hand, there seemed to be general agreement on the necessity for close contact and individual interaction during the development of the manufacturing, assembly, and testing procedures. Such close cooperation on the individual level was therefore encouraged and in large measure achieved during the training and start-up phases of the transfer process. Limited technical assistance was also supplied by Cummins personnel in Chongqing.

In developing the production process, Chongqing Automotive Engine Plant used local suppliers that were part of the system, that is, they came under the administration of the Sichuan Heavy Duty Truck Corporation. At the time, these suppliers were assigned to the Chongqing Automotive Engine Plant; more recently, however, limited freedom of choice was available.

MANAGEMENT OF THE ONGOING PROCESS

Communications and Feedback

All Cummins personnel who were interviewed agreed that communications between Cummins and the Chongqing Automotive Engine Plant were limited and that this circumstance resulted in inefficiencies in the technology transfer process. In their view, Chongqing Automotive Engine Plant limited the information transmitted and the prevalent attitude was: "You sent me the paper [i.e., technology documentation] and specifications, and now it's my business, not yours, what we do with it." They felt that they had bought the technology and this was the end of the process, except when they wished to ask questions. In the latter case, Chongqing Automotive Engine Plant personnel were quite forceful in demanding satisfaction, and in response to this requirement Cummins made important changes in the style of its responses over the course of the years. Specifically, the responses and explanations had become more detailed, obviously requiring a greater level of effort on the licensor's part.

The shortcomings in communication had, in the view of Cummins personnel, serious consequences for them. It was generally agreed that the information available to Cummins from Chongqing Automotive Engine Plant was inadequate for a complete evaluation of the Chinese product and the status of the technology transfer process. In fact, some at Cummins said that the company was still inadequately informed concerning the Chinese affiliate's absorption capacity for new technology, mainly because of limited access of Cummins personnel to facilities and engineers at the Chongqing plant. Also, the limitations imposed on open communications adversely affected the absorption of the dynamic technology by the Chongqing plant, it was held.

One Cummins engineer related his experiences during his many visits to the Chongqing Automotive Engine Plant. He said that on such occasions he had found the assistance of Chinese engineers who had been trainees at Cummins invaluable:

> Resident training at Cummins has been instrumental in [the Chongqing Automotive Engine Plant personnel] taking back with them some U.S. culture in technical and information areas. Typically, such trainees speak better English and [the Cummins visitor] tends to interact more with them during visits.

He also said that his own repeated visits had enabled him, with time, to build up a relationship with certain individuals. This was of great help in overcoming some of the barriers. The engineer described the typical consultation between him and the Chinese affiliate's personnel as taking place in a special reception room reserved for foreign visitors, far removed from the offices of the plant's staff. It then became a problem to locate needed background material, and this had to be brought into the meeting room piece by piece; there also was no table available for spreading out drawings, since the room was designed for receptions and formal exchanges. Only rarely did a Cummins engineer gain admittance to office facilities and then only for very short periods after which, so it seemed to him, he was expeditiously spirited out and taken back to the "building reserved for foreigners." The relationship with the Chongqing Automotive Engine Plant was characterized as "formal" and "stifling." Lack of information also prevented Cummins from giving support to Chongqing Automotive Engine Plant products in customers' hands in China, but then the Chinese affiliate was not interested in this type of cooperation anyway.

As one possible way of furthering communications, a permanent Cummins presence in China was considered. After some investigation into such a possibility in conjunction with and in support of the start-up of the diesel engine production at the Chongqing Automotive Engine Plant under Cummins' license, the U.S. company decided against such a step. The factors arguing against having an expatriate in residence in Chongqing included prohibitive costs and resistance by the Chinese affiliate to such a move. However, in retrospect some believed that a temporary presence of duration of about three months at the beginning of the start-up phase would have been decidedly beneficial. According to this view, a resident advisor would have been very helpful to the process of absorption of the Cummins documentation, but the Chongqing Automotive Engine

Plant was not willing to pay the necessary per diem expenses. It was also thought that it may have been beneficial for Cummins' broad scope of activities in China to have established a resident presence in China, for example, in Beijing, during the years 1981–1982 or immediately following the conclusion of the licensing agreement.[9] Such a representative could have been helpful in supporting the Chongqing start-up and in resolving at least some of the communication problems, as well as in furthering Cummins' direct marketing effort.

Manufacturing Planning and Performance

Planning and scheduling of production at the Chongqing Automotive Engine Plant suffered from the lack of availability of foreign exchange to the company. As a result, the Chinese plant could not order parts or kits before it had obtained a specific order for the product. Thus it could only order Cummins parts on demand and was constrained from establishing an inventory. To compound the problem, the Chinese affiliate was limited to the use of Chinese flag carriers, as specified by the agreement, resulting in further delays and a serious time lag between the receipt of the order and the delivery of the product. Cummins had very limited influence on this process and could only try to accelerate the shipping time of the kits from the U.K. plant to Shanghai. Possibly Cummins could have dealt with this problem in advance if it had been aware of it, but it had assumed that the Chinese plant would be able to obtain foreign exchange for activities that lay within the scope of the agreement. However, it later learned that Chinese governmental regulations applied to all transactions and could not be modified by contracts, even if these had been officially sanctioned; a Chinese company had to obtain an allocation of foreign exchange for every transaction. Marlow (1985) reported in his thesis that it took 10 to 15 months to deliver an engine to the customer from the time of the order being placed with the Chongqing Automotive Engine Plant. This compared with a delivery time of five months from the time of signing the contract for a direct purchase from Cummins.

Dennis Piper had been the primary engineering support person for the Chongqing Automotive Engine Plant licensing agreement since the inception of the start-up phase in early 1981. His responsibilities included monitoring and evaluating the performance of the Chinese affiliate and assessing its capability to integrate the transferred technology. However, Piper stressed that he did not evaluate the licensee in general, but only specifically with respect to the particular hardware being produced under Cummins' license.

The quality of the Chongqing product was considered "sufficient," and the output rate in 1985 was judged by Cummins as "reasonable." It was assumed by Cummins personnel that the Chinese product was up to standard in its quality

---

[9]The logistics of long-term residence for a foreigner are considerably more practical in Beijing or Shanghai than in most other locations in China.

since it was basically similar to the original product. The Chinese-made engines were largely assembled from Cummins parts and were tested in accordance with Cummins' specified procedures. The product had been "reviewed" by Cummins personnel, but the information available to Cummins was not sufficient to determine that the Chinese product "can be integrated in the original" Cummins product. According to experience with other licensees, "quality issues really arise when they start making large numbers of their own parts, and that was only just beginning [at the Chongqing Automotive Engine Plant]," the case study team was told in May 1985. However, the Engineering Department of the Chinese affiliate had not yet developed the capability to do the "engine specification," according to the judgment of Cummins engineers. Cummins was to provide this service for the Chinese affiliate until 1986, when it was to be discontinued. The current production volume of the Chongqing Automotive Engine Plant was about 700 units per year, but it should have reached double that figure, particularly in view of the countrywide demand, which was estimated at about 1,500 units.

Piper defined the criterion for the solution of any particular problem encountered by the Chinese affiliate, as the successful production of a document which they themselves could use. According to mutual agreement, the Chongqing Automotive Engine Plant had responsibility for producing that document. In the final analysis, the criterion for the complete absorption of the transferred technology was the feasibility of integrating the product of the technology recipient into Cummins' own product.

In accordance with the criteria defined above, the Chinese affiliate, Chongqing Automotive Engine Plant, had not yet reached the stage of product specification capability, nor had it reached the stage of complete technology assimilation. However, the Chongqing Automotive Engine Plant had successfully installed its assembly line and test facility. At the time of the interviews, the plant produced NH-series engines, and it had successfully assembled its first K-series engines. The local added value amounted to about 30 percent, according to the Chinese partner. As already discussed, Cummins had difficulty verifying the status of the technology transfer process accurately due to the limitations imposed by the Chinese partner on communication between the companies.

Cummins' managers were of the opinion that the Chongqing Automotive Engine Plant management was good and as a result the joint project had progressed at a reasonable rate from 1983 to 1985. "The new management was on a very different course," compared with its predecessors, it was held. Also, substantial progress had been made in physical facilities; new factory structures had been designed and built and test cells had been constructed.

Customer Service

Cummins staff performed marketing services for the Chinese affiliate during the years 1981–1982, but later the Chongqing Automotive Engine Plant found its own customers and became quite independent in this area. The Chongqing Automotive Engine Plant sold its product under warranty and serviced it for its customers; it was considered to have made much progress since 1981 in the area of customer relations. Cummins feels that it should have participated in the maintenance of the licensee's product in the field, but, as already mentioned, the Chinese factory did not wish to cooperate with Cummins this closely on a continuing basis and had not been responsive to Cummins' proposals in this direction.

Customer Relations for Direct Marketing

The Cummins China Business Group was responsible for the direct marketing of the company's product in China. As it turned out, direct sales constituted an important component of the China business activities and included engines identical to the Chongqing product. According to the licensing agreement, the Chongqing plant did not enjoy exclusive rights to the China market, and Cummins continued marketing its U.S.-made product. This activity led, in some instances, to competition between the licensor and the licensee.

It appeared that many Chinese customers wished to buy the original product directly from the U.S. company because of the adverse reputation the Chongqing Automotive Engine Plant acquired over the years through its sales of the French Berliez diesel engine for which the license has been acquired in the 1960s. This engine was still in production at the plant and had a history of performance problems.[10] In spite of substantial improvement in the management in recent years, the reputation of the Chongqing plant still suffered from the general widespread dissatisfaction with the Berliez engine; this circumstance adversely affected the sales of the Cummins engine. For example, the Pingshuo Coal Mine project, a joint venture between Occidental Petroleum Corp. and the Ministry of Coal Industry, decided to buy engines for their operation of off-highway trucks directly from Cummins. Also, three bulldozer plants bought diesel engines to power this equipment partly from Cummins and partly from the Chongqing Automotive Engine Plant, thus compounding the maintenance problem by requiring separate responsibility for the identical engines. In the opinion of U.S. personnel, future Cummins activities in China may be expected to cause still more confusion.

In one case, both Cummins and the Chongqing Automotive Engine Plant bid for the same sales contract, and Cummins was in the embarrassing position to have underbid the Chongqing plant. This situation came about because Cummins was competing also against Caterpillar and General Motors and had to lower its price to win the bid. At the same time, this situation could indicate that Cummins parts were sold to the Chongqing Automotive Engine Plant at higher than necessary prices. In hindsight, Cummins may have made an error

---

[10]All parts and materials for the Berliet engine were sourced in China.

in judgment in this regard. It was learned that the incident described above was not an isolated case and the problem of competitive pricing between the two affiliates was expected to persist.

The servicing and customer follow-up for the direct sales became in 1984 the responsibility of the Technical Service Center, which was established in Beijing as a cooperative venture between Cummins and the Ministry of Machine Building Industry (Marlow, 1985). This Center was originally conceived to do all servicing, including sales of the Chongqing affiliate, but the latter resisted cooperation in this area.

Cummins had hoped to set up several additional service centers in different locations where engines, sold directly to the end user in China by the U.S. company, were in use. These plans met with resistance from the Chinese bureaucracy, which limited Cummins to a single facility. On the other hand, the administrative barriers between different ministries made it difficult, if not impossible, to service a wide variety of direct customers from one service center affiliated with a specific ministry. The ministries primarily concerned included the Ministry of Metallurgy, which operated the major mineral mines, and the Ministry of Coal Industry, which administered China's coal mines. It was hoped that this impasse would be overcome in the near future, but, meanwhile, the maintenance of the equipment in China was a considerable burden in view of the considerable expense involved in supporting an expatriate staff. Additional problems had surfaced in conjunction with the administration of the Technical Service Center in the area of staffing and creating work incentives.

Assessment of the Current Position

The lack of satisfactory communications between the two companies limited the capability of Cummins to accurately assess the current position of its cooperation with the Chongqing Automotive Engine Plant. Although the output of the Chongqing plant was considered to be reasonable, it could double its present production level and still not exceed the demand of the Chinese market, according to some Cummins managers. Concerning the quality of the product, Cummins personnel were guarded in their assessment, again because of incomplete data, but they generally believed that the product was satisfactory. Cummins' sources agreed that the present Chongqing Automotive Engine Plant management was performing well; however, all Cummins personnel involved in the cooperation also were of the opinion that the Chinese affiliate still kept its distance from the U.S. licensor and that information flow was still far from satisfactory. Cummins' management assessed the Chongqing experience as positive and profitable for the company. Thomas Head expressed it this way: "In Chongqing, what has happened, and everyone knows it, was that we were ahead of the game." Cummins sold engine kits and parts for hard cash and, in addition, had developed a market for direct sales in China which brought in increasing returns.

FUTURE PLANS

Further Ventures

Technical Service Center. In July 1983 Cummins signed an agreement with Techimport and the China Automotive Industry Corporation (CAIC) now referred to as CNAIC for the establishment of a Technical Service Center in Beijing. Marlow in his thesis (1985) devoted a chapter to a description of the negotiations and the functions of this Center. The great majority of the information given below was taken from this source. The Center was Chinese owned and managed with Cummins having responsibility for the training of its personnel in the areas of administration, servicing and repair of Cummins engines, applications engineering, and marketing. The Technical Service Center was managed by Heavy Duty Associated Motor Vehicle Company, a subsidiary of CAIC, and the parent corporation, as a signatory to the agreement, has the responsibility for the buildings and staffing of the Center. Cummins has established a permanent presence in Beijing in the form of a Regional Office with a staff of two. Stephen Mulder, former manager of Marketing of the China Business Group, was appointed regional manager for China and took up residence in Beijing in June 1984. Wei-ming Tan, a U.S.-educated Cummins engineer who emigrated from China to the United States in 1981 after having lived there for 25 years, moved to Beijing in the spring of that year. Tan's responsibilities centered on applications engineering, maintenance, and repair. Through this Regional Office Cummins provided technical and administrative assistance to the Technical Service Center. The Center had a staff of 15, including two who had special political supervisory functions. Techimport, which was the negotiating agency of the Ministry of Foreign Economic Relations and Trade (MOFERT), became an active partner to the operation, as in the case of the license agreement with the Chongqing Automotive Engine Plant. Techimport was responsible for supervision of the Technical Service Center for the central government and, in particular, was responsible for all foreign currency operations of the Center. The Technical Service Center was partly financed by commissions paid by Cummins for the sale of engines and parts sold in China through the marketing efforts of the Center. These commissions constituted foreign currency income and were transmitted through Techimport to the Ministry of Finance. It was originally hoped that the Technical Service Center would be allowed to retain a part of these commissions to use for work incentive programs. However, this hope was not fulfilled and, as a result, a lack of incentives for the staff caused lower productivity levels than projected.

The Technical Service Center agreement was believed to be a landmark in a number of respects since it deviated from previous servicing agreements entered into by Chinese agencies with other foreign companies.

1. The agreement was nonexclusive. Cummins insisted on this condition since it realized that it would be necessary to establish a network of service sta-

tions, at least one for each ministerial aegis under which the various end users operated. This necessity arose from the administrative barriers between Chinese ministries.

2. The agreement was comprehensive in nature, including the functions of sales promotion for Cummins products, applications engineering, servicing and service training in the field, in addition to the more accepted functions of maintaining of repair facilities and parts distribution.

3. Cummins' contribution to the Center's revenue was in the form of commissions based on the sale of engines and parts rather than a fixed annual fee as originally proposed by the Chinese side.

The agreement for the Technical Service Center was concluded in a relatively short time. Originally, another agency of the Ministry of Foreign Trade, Machimpex (China National Machinery Import and Export Corporation), was designated to negotiate the agreement with Cummins since it had negotiated and concluded previous agreements of similar nature with foreign companies. However, Machimpex negotiators proved to be inflexible and insisted on Cummins accepting terms similar to those contained in previous agreements. Cummins decided on a hard line and informed its counterparts that it would not participate in any further negotiations with Machimpex. This was in December 1982, and in April 1983 Cummins was informed that negotiating responsibility had been transferred to Techimport, which was the agency preferred by the U.S. company. It must be added here that the Chinese prospective implementer of the agreement, CAIC, had active interest in the establishment of the Technical Service Center and was believed to have lobbied forcefully for the change of negotiating agency. The agreement was signed in July 1983, only three months after the initiation of the negotiations with Techimport. Cummins personnel felt that the negotiations proceeded with such efficiency partly because of the growing mutual confidence between the negotiating partners based on their previous interaction and partly because the company had learned a great deal from its earlier negotiating experience and had a better feel for the timeliness of taking a hard stand. Dennis Kelley conducted most of the negotiations for the Service Center and was on the scene when the decision was taken to abandon negotiations with Machimpex.

Other future ventures.   In late 1985 Cummins was investing considerable amounts of personnel time on "courtship" of Chinese factories, "listening and planning for their needs and adapting to the country's conditions," according to Charles Byers, vice president for International Business Development. Byers argued that China was not an attractive consumer market because of the low per capita income, but it was an important and expanding market for industrial goods. In the development of China, "portable power was basic," he said, and the future challenge for Cummins, in his view, lay in the on-highway sector. The company had established a position in the high-power (250-horsepower and up) off-highway vehicle market, and most of the Cummins motors in China were used to power trucks at mining sites. According to Byers, the Chinese government had already made important decisions aimed at developing transportation by on-highway trucks, and lower power diesel engines (250-horsepower and lower) were required in large quantities to power these trucks. A cooperative venture with the No. 2 Automotive Works in Shiyan, Hubei Province, was a possibility. "There was no doubt concerning the capability and technical competence" of this plant, according to Byers. He added, "It has the best metal cutters in China and has acquired a new foundry from Germany." Byers also said that this plant had a capacity of 40,000–50,000 units per year, much larger than that of the Chongqing Automotive Engine Plant, whose capacity he estimated at 1,200–2,000 engines. Another factory of interest to Cummins was the Shanghai Diesel Engine Plant, which was subordinate to the Ministry of Agriculture and whose mission therefore was the support of agricultural development.

Cummins evidently had given much thought to the development of mechanisms that would significantly expand its activities in China. The company had concluded that all cooperative commercial ventures in China were limited by the availability of foreign exchange and it was therefore of primary importance to find methods to generate foreign currency income. One method under consideration was referred to as *trading*: exchanging Cummins products for any Chinese product marketable abroad. The idea was akin to the "packages" reportedly developed by many Japanese companies by which two or more businesses join together, some of them exporting to China and others importing from China, thus eliminating the necessity for the Chinese buyer to obtain foreign exchange. A second possibility was the development of Chinese industries as low-cost sources of parts and components. Byers said that Cummins was committed to the upgrading of local Chinese technology, thereby facilitating this type of interaction. The third possible method of cooperation was licensing under agreements similar to the one already concluded for the production of engines by the Chongqing Automotive Engine Plant.

On the other hand, it was not clear whether conditions in China were favorable for Cummins to enter into joint venture equity agreements. For example, a patent law had only recently been implemented but remained largely untested. Also, joint ventures had, by current law, fixed terms, although the limit had been extended to 50 years. Cummins was also not clear on the validity of profit computation, given the absence of an open market and the resulting prices, which were, in the view of the company, artificially fixed. It should also be added here that some disagreement was noted concerning the future promise of China as a low-cost source; Thomas Head believed that the savings arising from the development of advanced automation in industrialized countries would overtake the economic advantage of low labor costs in developing countries within about five years. However, the pressure to raise foreign currency "may well cause a push to use China as a cheap source of components and of equipment," and further, under the given circumstances, "the push may go beyond hard economic reason," which would, in fact, amount to subsidies for the project concerned, according to Head.

Cummins planned to conclude negotiations in the immediate future on additional service facilities for the company's product installed in equipment used by Chinese enterprises subordinate to other ministries than the MMBI. Cummins was optimistic that bureaucratic obstacles to the establishment of such service and parts distribution centers could be overcome and that the service centers could be established during the next three to four years. Cummins also hoped that future agreements could be concluded for such facilities after the model of the Beijing Technical Service Center. It was felt that the development of new cooperative production projects would significantly exacerbate the increasingly confused situation in the area of customer service for Cummins products in China, including both locally and foreign produced.[11] Thus, the development of an adequate network of service facilities required the urgent attention of the company's management.

It was learned that Cummins wished to make a number of modifications in future agreements relative to the licensing agreement for the production at the Chongqing Automotive Engine Plant. For one, Cummins would seek prepayment for the contracted personnel training programs. This condition would address the shortfall of personnel sent for training by the Chinese entity, presumably because of its desire to save foreign exchange. Another important issue to be resolved in future agreements was the problem of competition between Cummins' direct marketing efforts and those of the licensed affiliates. The company became sensitized to this problem by its experience with the Chongqing Automotive Engine Plant venture. In future agreements, it was held that Cummins must think carefully about pricing and establishing customer confidence in the licensee's product. In the case of the Chongqing plant agreement, Cummins originally attempted to market the affiliate's product, but it was now clear that Cummins had not been successful in this effort. It was nevertheless believed that Cummins had helped and supported Chongqing Automotive Engine Plant in its market development and that the Chinese affiliate had derived advantage from Cummins' sale of components in China.

In addition to the search for further suitable affiliations with Chinese companies and corporations, Cummins planned to continue direct marketing operations in China, according to the company's executives. In fact, the company hoped to utilize service facilities as agents in these efforts as they had done successfully with the Technical Service Center.

Dennis Piper discussed the problems encountered by Cummins in the course of technology transfer activities during implementation of dynamic licensing agreements with enterprises in developing countries. He pointed out that Cummins' technology moved very rapidly, perhaps too rapidly for most af-

---

[11]There were over 1,000 Cummins engines in use in China in 1984 (Marlow, 1985).

filiates to be efficiently updated continuously. He said that it might be desirable for Cummins to slow down its development and updating process to better accommodate foreign affiliates. Another alternative would be to prepare "a box of specifications" at two-year intervals, for instance, rather than supplying a continuous flow of innovation, Piper said. These questions concerning modifications to Cummins' procedures to better accommodate foreign affiliates were currently under discussion.

According to Thomas Head, Cummins did not envision future cooperation with Chinese enterprises to include design of new lines of products. Rather, he saw collaboration to emphasize "applications engineering" or the adaptation of Cummins products to specific equipment produced by Chinese or other OEMs (original equipment manufacturers). On the other hand, Head conjectured that, eventually, limited joint design projects may be pursued; such projects had been initiated with the company's affiliate in India.

EXHIBIT 1 Cummins Engine Company–Chongqing Automotive Engine Plant:
Key Events

Date	Event
1978	
August	First meeting in Beijing
September	Cummins' team of manufacturing experts visits China
December	Second negotiation meeting, in Beijing
1979	
March	First Cummins contract draft sent to China
April	Third negotiation meeting in Beijing
December	Fourth negotiation meeting in Beijing; Cummins' first price proposal presented
1980	
March	Fifth negotiation meeting in Beijing; revised price proposal presented
July–August	Sixth negotiation meeting in Beijing; Cummins' delegation leaves meeting
November–December	Seventh negotiation meeting in Beijing
1981	
January	Eighth negotiation meeting in Beijing; licensing agreement signed
February	Cummins' China Business Group established
March	Export license issued by U.S. Department of Commerce
June	Cummins visits Chongqing Automotive Engine Plant to determine feasibility of manufacturing both N- and K-series engines at this plant
September–November	Joint Planning Session in Columbus, Indiana
1982	
January	First group of Chinese trainees arrive in Columbus, Indiana
September	First shipment of Cummins engine kits arrive in Chongqing
1983	
July	Technical Service Center Agreement signed in Beijing
1984	
June	Cummins' regional office in Beijing opened

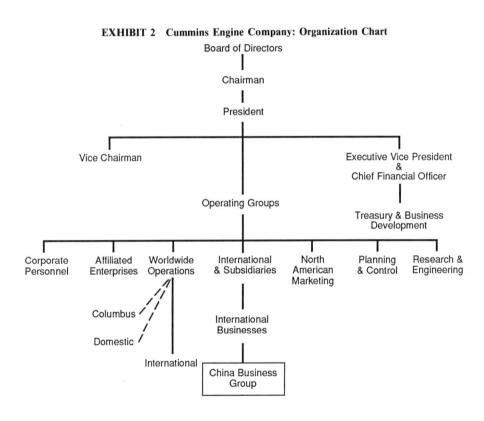

EXHIBIT 2 Cummins Engine Company: Organization Chart

Board of Directors

Chairman

President

Vice Chairman

Executive Vice President
&
Chief Financial Officer

Operating Groups

Treasury & Business
Development

Corporate
Personnel

Affiliated
Enterprises

Worldwide
Operations

International
& Subsidiaries

North
American
Marketing

Planning
& Control

Research &
Engineering

Columbus

Domestic

International

International
Businesses

China Business
Group

EXHIBIT 3   China Business Group: Organization Chart

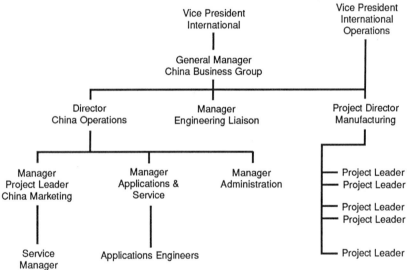

Vice President
International

Vice President
International
Operations

General Manager
China Business Group

Director
China Operations

Manager
Engineering Liaison

Project Director
Manufacturing

Manager
Project Leader
China Marketing

Manager
Applications &
Service

Manager
Administration

Project Leader
Project Leader

Project Leader
Project Leader

Project Leader

Service
Manager

Applications Engineers

CASE 6

Cummins Engine Company: The Chinese Perspective

INTRODUCTION

This case was based on an interview trip to China in June 1986. The research team visited the Ministry of Machine Building Industry (MMBI) (now called Machine Building Industry Commission) in Beijing and the Chongqing Automotive Engine Plant in Chongqing. The purpose of these visits was to investigate the Chinese view of the licensing agreement concluded in January 1981 between the Cummins Engine Company in Columbus, Indiana, and the China National Technical Import Corporation (Techimport). The Chongqing Automotive Engine Plant was designated as the recipient of the Cummins diesel engine technology to be transferred to China under the agreement. At the MMBI, discussions were held with representatives of the China National Automotive Industries Corporation (CNAIC) and of other bureaus.

The MMBI was one of about 50 ministries, commissions, and other high-level government agencies, whose heads, the ministers, made up the State Council of the People's Republic of China. The State Council was headed by Premier Zhao Ziyang and four vice premiers. Each ministry had countrywide responsibility for the global planning and for the execution of large-scale projects in a particular sector of the economy or a service area. Examples are the Ministry of Metallurgical Industry, Ministry of Petroleum Industry, Ministry of Labor, State Planning Commission, State Science and Technology Commission, and so forth.

169

The bureaus of the MMBI included the China National Automotive Industry Corporation (CNAIC), formerly known as the Automotive Bureau; Instrumentation Industry Bureau; Electrical Equipment Bureau; Machine Tool Bureau; Heavy Machinery and Mining Equipment Bureau; Basic Components and Parts Bureau; Agricultural Equipment Bureau, and Science and Technology Bureau. The Ministry also maintained a Science and Technology Information Center. The research team met with representatives of all the bureaus mentioned above in a general session at which the U.S. visitors presented an outline of the research project comprising the case studies of technology transfer projects of Cummins Engine Company, Foxboro Company, Westinghouse Electric Corporation, and Combustion Engineering. These U.S companies' projects were the responsibilities of CNAIC (Cummins Engine Company), the Instrumentation Industry Bureau (Foxboro Company), and the Electrical Equipment Bureau (Westinghouse Electric Corporation and Combustion Engineering). The team subsequently met with representatives of each of these bureaus separately to obtain information on their experiences with and their specific attitudes toward each of the projects.

The following ministries and commissions would typically be involved in the process of negotiation, approval, and implementation of a technology transfer agreement with a foreign company. The ministry of the relevant industrial sector, through one of its bureaus, would develop the project to the negotiation stage. In previous years, one of the foreign trade corporations under the Ministry of Foreign Economic Relations and Trade (MOFERT) was then designated to take charge of the negotiations. Techimport, the agency which negotiated the Cummins Engine Company diesel engine licensing agreement, was one such corporation. In recent years, at least some of the ministries have been authorized to establish their own foreign trade corporations to improve the efficiency of the negotiating process.[1] Once agreement had been reached on this level, the industrial ministry (MMBI in the present case of interest) carried the approval process forward to the State Council, MMBI sources have said. Usually, the State Economic Commission and the State Planning Commission signed off on the agreement before the State Council granted final approval.

DEVELOPMENT

MMBI maintained a wide range of contacts and was in communication with many companies all over the world. In the recollection of ministry personnel, U.S.-China technology transfer began a long time ago; the earliest affiliation was with Westinghouse in 1946.

---

[1]However, MOFERT still remained involved and had to approve the commercial terms of the agreement.

The need for more advanced diesel engine technology arose in several sectors in China, including the powering of off-road vehicles used in construction and mining, particularly for coal mining, as well as on-road transportation.

The Chongqing Automotive Engine Plant with 2,600 employees was said to have had a good basis of diesel engine technology.[2] In 1965 the factory had acquired foreign technology from a French company, Berliez, but production at the plant was interrupted during the Cultural Revolution (1966–1976). In 1978, the Chinese government decided to develop and modernize the Chongqing Automotive Engine Plant and to import modern diesel engine technology through a licensing process.

The fact that Cummins-produced engines were used by many truck manufacturers whose products China imported, influenced the decision to select Cummins as the desired source of diesel engine technology.[3] In addition, the Japanese company, Komatsu, from whom China contracted for the production of bulldozers under license, used Cummins' technology to power its own product. Since Komatsu by its agreement with Cummins was not permitted to sublicense Cummins' technology, the choice of Cummins by the Chinese authorities was furthered by the need for engines compatible with this equipment. Engines of 220 to 360 horsepower were required in this case, but also engines of larger capacity (up to over 1,000 horsepower) were of interest for other applications.

In the years 1978, 1979, and 1980, Cummins personnel visited the Chongqing plant and concluded that the factory's conditions were suitable for an affiliation. Also, local suppliers of the plant were judged to be at a suitable technical level. In return, a delegation from MMBI's Automotive Bureau visited Cummins Engine Company in 1979 in order to investigate the U.S. company's products and technology.

China-U.S. Interaction

MMBI officials said that U.S.-China technology transfer had a number of inherent problems, arising from the basic differences in the social and cultural systems. The differences in the customary infrastructure of the industrial and commercial organizations in the two countries hampered interactions among companies of different nationalities, they added.

According to Song Juzhi, China did not always seek to acquire technology from its original developer. He said that there were some examples of U.S. technology that was transferred to China from Japan. Sometimes, the original innovator lost the lead position to the affiliate who developed the technology

---

[2]In the Chinese context, the staff included many who were employed in services outside the main mission of the company, including teachers, medical staff, and others.

[3]China imported Wabco, Terex, and Mack trucks of capacities 108 tons, 152 tons, and also some lighter trucks.

further and then it became advantageous to acquire it from the latter. In addition, cost advantages may also have favored the secondary source of the technology. Also, standards of materials in the country of secondary origin may have been more closely similar to those in use in China than those in the country in which the technology was first developed. All these conditions have in the past contributed to decisions in favor of the secondary source. Additional motivation had to exist for the decision to deal with the primary developer of a technology.

Song Juzhi summed up the conditions which, in his view, strongly influenced the interaction between the negotiating teams of China and of a foreign company when technology transfer was under consideration:

1. The Chinese and foreign companies should be familiar with one another. The greater the familiarity, the easier the negotiations.
2. The Chinese commercial term negotiators had to become more knowledgeable than they currently were. Better training in this regard would expedite the negotiating process.
3. The level of risk involved in a project greatly influenced the time it took to conclude an agreement. The greater the risk, the longer the negotiations process, in the experience of the Chinese officials.
4. The type of personnel involved determined the level of difficulty of the interaction. Technical personnel interacted easiest. Commercial personnel experienced greater difficulties. American lawyers were the most difficult to deal with. (Cummins did not send lawyers to join their negotiating team; they only sent a contract specialist.)
5. The English and U.S. system of measurements, technical conventions, and standards of materials are inconvenient for Chinese engineering and manufacturing. For this reason, European and Japanese technology suppliers have a competitive advantage in China.

NEGOTIATIONS

Negotiations between Cummins and Techimport began in August 1978, and the agreement was signed in January 1981. The negotiations thus extended over two and a half years and included eight rounds. The time required was so long because the negotiations began soon after China opened up to the outside world following the period of isolation during the Cultural Revolution. As a result, Chinese personnel were not familiar with international commercial practices. The Cummins' personnel worked a total of 57 person-trips, totaling 906 person-days. Chongqing plant managers said that the process of negotiation has since been made more efficient as a result of a number of reforms on the Chinese side and accumulated experience on both sides.

Negotiations began with the choice of technology to be transferred. The commercial terms were only considered later. This order was the usual practice, according to the MMBI officials who met with the research team. During the first stage of negotiations, personnel of Cummins Engine Company introduced the Chinese representatives to the company's products and technology and described its production processes. The material presented was only brief and not detailed enough to satisfy the Chinese engineers, according to Chongqing plant managers. As a result, a delegation of MMBI's Automotive Bureau traveled to the United States and visited Cummins in 1979 in order to broaden the Chinese side's knowledge of their prospective affiliate.

The technology of interest in the project included the NH-series engines, which have 6 cylinders and range from 230 to 400 horsepower, and the K-series engines, which have 6, 12, and 16 cylinders ranging from 400 to 1,800 horsepower. In order to implement the licensing agreement, Chongqing Automotive Engine Plant required extensive documentation including blueprints, technical standards, and descriptions of patented technology. At first, it was not clear to the Chinese negotiators that Cummins could not supply such documentation for components sourced from outside the company. However, this circumstance was finally clarified and understood by the Chinese side before the agreement was signed. Subsequently, and as a result of a negotiated understanding between the parties, the Chongqing plant was able to establish contact with Cummins' U.S. suppliers.

The Chinese side wished to expand the agreement and to obtain information concerning diesel engines outside the scope of the central project. However, Cummins remained adamant and strictly limited the technology to be transferred to that required for the manufacture of the product lines included in the agreement, according to information obtained from Chongqing plant managers. The Chinese side was, however, pleased by the Cummins preference for dynamic (as opposed to static) licensing, since this was also their first choice.

Negotiation of the commercial terms of the technology transfer agreement was conducted after agreement had been reached on the technical aspects. Licensing was the only process considered, since, from the Chinese point of view, it was too early in the "opening-up" period for the government to approve foreign investment. Cummins preferred a one-time payment, but the Chinese side considered this as placing an unfair share of risk on their side. Agreement was finally reached on a system of royalties based on the value of both product and spare parts.

According to MMBI sources, the parties to the negotiations faced a crisis concerning pricing. The Chinese side considered the prices charged by Cummins to be too high compared with other commercial ventures in their experience and also compared with prices quoted by Cummins' competitors. In the end, Cummins agreed to lower its prices. The research team could not obtain additional details.

Cummins impressed the Chinese negotiators as being "quite open and non-conservative" in comparison with other foreign companies with whom the MMBI officials had had contact. For example, Cummins' negotiators pressed for more training in terms of man-months than the Chinese side proposed. According to previous experience, other foreign companies wished to limit the scope of training for Chinese personnel.

MMBI officials said that, in their opinion, some of the terms and conditions included in the final licensing agreement with Cummins were "not reasonable." Regional restrictions were imposed on the export of engines produced in Chongqing under license. Also, requests for the CKD (completely knocked down) and SKD (semi-knocked down) units bought for assembly at the Chinese plant had to be channeled through Cummins Engine Company, even though they were supplied by an affiliate. Similarly, the Chinese company was bound by the agreement to purchase all parts and components through Cummins, even if they were produced by other suppliers. In the latter case, Chongqing knew who the suppliers were but was constrained from dealing with them directly. In some cases, Cummins charged higher prices than the original manufacturer for direct purchases. These complaints were voiced by MMBI personnel, but the managers of the Chongqing Automotive Engine Plant who were interviewed did not mention them.

It was estimated that in 1986 it would take about one-quarter of the time to conclude a similar agreement with Cummins or another company with whom the Chinese side was already familiar. MMBI officials said that the reasons for the improvement of the efficiency of the negotiating process were several. The Chinese side was now more familiar with the terms and conditions of international contracts as well as with the whole negotiating process. Also, they said that they were more familiar with Cummins and were therefore able to interact with the U.S. personnel with greater confidence. After years of interaction with the U.S. company, including training and consultation, as well as through exposure to Cummins' literature, the Chinese engineers had also come to know the Cummins product line, further facilitating the process. The team was further told that CNAIC had acquired the authority to negotiate directly, which further increased efficiency. As previously noted, only the foreign trade corporations of MOFERT had such authority earlier on, and CNAIC had to negotiate with Cummins through the intermediary, Techimport. Managers of the Chongqing Automotive Engine Plant said in June 1986 that the plant had also, by then, been given the authority to negotiate directly with foreign companies. This procedure allowed for a direct connection between the foreign supplier and the Chinese recipient of the technology.

MMBI officials shared some information concerning their negotiating practice and strategy. Discussion of cost was considered to be of the greatest importance, and the usual strategy pursued was as follows. The Chinese negotiators attempted to obtain as much information as possible from outside sources concerning the cost of the technology and the products to be acquired.

They were then in a position to hold foreign price demands to reasonable levels. It was found most effective if they could quote the prices of the foreign company's competitors. However, world market prices were not always widely known, limiting the applicability of this tactic. The higher the first quote offered by the foreign company, the longer the negotiations on this subject have taken, the team was told. On the other hand, in the view of the MMBI officials, the U.S. side often inflated the prices in the initial quotation because they knew that the Chinese were bargainers. As a result, negotiations were often prolonged since in the end reasonable prices had to be arrived at before an agreement could be concluded.

START-UP

General

Detailed plans for the start-up process were approved at the Joint Planning Session, which took place in the United States in September through November 1981. In anticipation, the Chongqing Automotive Engine Plant began to implement internal reforms in the areas of management and quality control in order to adapt to the Cummins technology transfer process, since the original structure of the Chinese company was too different from that of the U.S. company. In spite of this effort, many problems were encountered. Chinese standards for preparing engineering drawings are different from U.S. methods, and of course all measurements had to be translated into the metric system of units. In addition, standards of materials in China do not conform to the U.S. system and had to be translated into Chinese terms.

Documentation

Documentation was delivered to China within the schedule limits set forth in the agreement. Minor deficiencies were encountered: some blueprints were incomplete, and some pages were found missing. However, these inadequacies were promptly and effectively corrected by Cummins. In fact, the agreement provided for fines in case of such deficiencies, but this clause was never implemented.

The greatest problem encountered in this project was the absence of subassembly drawings which Chinese manufacturing engineers rely on for gaining understanding of relationships between individual parts. Only drawings of the individual parts were supplied. As a result, design engineers at the Chongqing plant had to prepare the needed assembly drawings, thus delaying start-up of the manufacturing process by several months. However, more important, the incident was a setback in the relationship between the companies, since the Chinese side was deeply disturbed by it; they at first believed that Cummins had willfully withheld these drawings. Only after a visit to the U.S. plant did the Chinese engineers convince themselves that Cummins did not use intermediate assembly drawings, similar to those routinely used in China. It turned out that "assembly routines" or assembly instruction documentation were used at Cummins' plants, and these were also supplied to the Chinese affiliate. Liu Xianzeng (senior engineer, CNAIC) said that the assembly routines could be used for assembly and manufacturing but were not adequate for design work. This difficulty in the interaction was still very much alive in the minds of the MMBI officials, but the incident was never mentioned during the Chongqing Automotive Engine Plant visit. On the positive side, Liu added that Cummins supplied microfilm records and computer software, some of which was relevant to the history of the engines' development. This material was found to be very valuable.

Training

In the opinion of MMBI officials, Cummins was quite open and cooperative and its personnel were hard working and enthusiastic in implementing the training program under the licensing agreement. The agreement provided for training amounting to up to 600 man-months, of which more than 100 man-months had been utilized by June 1986. Of this training, more than 30 man-months had been used by suppliers of the Chongqing plant. The trainees typically spent four to five months in the United States. A good number of trainees worked at the Cummins Technical Center, where they have been engaged in joint product development projects, limited however to the NH- and K-series engine technology.

As the implementation phase progressed, the Chongqing plant required further support by Cummins, and it was important to supervise the start-up process and to monitor progress. Cummins sent competent consulting personnel to Chongqing during the start-up phase, and these U.S. personnel proved to be effective and cooperative. Consultations in China amounted to 130 man-visits. The two parties held meetings once a year during which progress was reviewed and procedures adjusted. Chongqing bore sole responsibility for developing the manufacturing process. Locally sourced components and subassemblies had to be tested to qualify them, and at Cummins' insistence U.S. personnel were present during these procedures. Cummins also assisted with advice and support in the purchase of foreign-made production and testing equipment.

In general, Chongqing plant managers were satisfied with the training program provided by Cummins, which was characterized as effective. Mostly engineers and management personnel were sent to the United States for training, and only a few workers participated in training overseas. The Chinese personnel were able to observe production procedures, and they could participate in production in plants where there was no union shop. In some U.S. factories the trainees were permitted to take photographs as well. The trainees encountered relatively few problems and "certainly had less difficulty than Japanese personnel," the researcher was told. The reference was to the resistance of American workers to passing on expertise to Japanese trainees, whom they view as competitors. When problems did arise, they were corrected following appeals to Cummins' project coordinators.

All Chinese personnel sent abroad for training followed preplanned programs that had been worked out in detail by the two parties to the agreement. Following the return of trainees to China, the Chongqing plant organized internal training programs led by the personnel who had been at Cummins' facilities, using their training plans as the basis for the discussion sessions.

MANAGEMENT OF THE ON-GOING PROCESS

Continuing Cummins' Support

Cummins' support to its Chinese affiliate extended to many areas, including production, quality control, customer service, and application engineering and design. Problems encountered in the course of the relationship centered around defective CKD and SKD units supplied by Cummins' British affiliate. This difficulty caused production delays and even stoppages at the Chongqing plant. However, Cummins Engine Company made great efforts to deal with these incidents to the extent of sending a corporate vice president to China. The research team gained the impression that this type of problem was now behind the affiliates, although it was considered serious at the time.

Schedules

Supplies from abroad were ordered one year in advance, and Chongqing Automotive Engine Plant usually kept an inventory of kits and other special parts sufficient for one month. The accepted time of delivery, that is, the time elapsed between the customer's order and delivery of the engine, was about one year; it took half a year to actually build the engine. Output of the Chongqing plant had reached a cumulative total of 2,000 units by June 1986.

Quality of Product

The Chinese customers of the Chongqing plant were mostly satisfied; the quality of the product had reached the U.S. company's standards. The Chinese factory had seven applications engineers who worked with customers on custom design and application modification of the diesel engines. Chongqing was independent in its customer maintenance and service operation. The Technical Service Center established by CNAIC in cooperation with Cummins only served the direct sales of the U.S. company in China.

Localization of Product

Local sourcing has progressed steadily since the beginning of production in Chongqing. According to the contract, all substitutions were subject to approval by Cummins. At the insistence of the U.S. company, qualification of locally sourced parts was conditional on testing performed in Chongqing with the participation of, or at least in the presence of, Cummins engineers. Major differences between standards for materials and components in China and those in the United States, Chinese executives explained, necessitated the determination of mechanical properties such as tensile strength and modulus and comparison with U.S. specifications supplied by Cummins. Chinese managers said that, in principle, all components could be produced locally, but the level of relevant metallurgical technology for and the resulting reliability of some selected components (e.g., main bearings and crank shaft forgings) had not yet reached the required level. Nevertheless, by June 1986 local content of NH-series engines produced in Chongqing had reached 80 percent of total value, while the local content of K-series engines was 17 percent.[4] Five years from the beginning of the project imported content was to be limited to selected components, as noted above.[5]

Competition and Finances

Cummins was permitted to market directly in China under the agreement. However, the Chinese executives said that they were not concerned about competition between Cummins and the Chongqing plant. In fact, the Chinese plant's output could not yet meet the demands of the local market, and it was the view of the Chongqing management that Cummins' marketing activity in China helped to develop the domestic market for its own product. The Chongqing plant's price for an engine was about 20 percent below that of Cummins' FOB charge, and the Chinese customer paid in local currency for the Chongqing product.[6] The Chinese factory required some foreign currency to cover the cost of the imported components, but it had had no difficulty obtaining the necessary allocations; it enjoyed favored status as an awardee of a national prize for its successful absorption of foreign technology.[7] As a result, foreign exchange did not present the major problem in this project that it often has in cooperative

---

[4]Production of K-series engines at the Chongqing Automotive Engine Plant began only in 1985.

[5]Since the agreement was signed in January 1981 and production began later in the same year, five years later would put the date of reaching this level of localization in the latter part of 1986.

[6]However, the Chongqing Automotive Engine Plant's product was "much more expensive" than domestic technology engines, the researchers were told. It was learned from a Cummins source that the cost ratio was close to 3.

[7]According to Cummins sources, this award provided for a 30 percent salary bonus for all employees for one year.

ventures with foreign companies. The national award was evidence of the recognition the plant had gained in China for its high standards of management and product quality, the Chongqing managers said.

FUTURE PLANS

In the view of both MMBI officials and Chongqing plant personnel, cooperation and communication between the parties to the agreement have been good; cooperation with Cummins was "smooth and successful," according to Liu Xianzeng.

MMBI officials said that plans for the immediate future called for increased scope of cooperation with Cummins in the form of additional licensing agreements, rather than joint ventures. Chongqing plant managers said that they hoped to discuss plans for extended activities with Cummins executives during a planned meeting in Chongqing in June 1986. They hoped to discuss possibilities for co-production or compensation trade, including the sale of locally produced parts to Cummins, thereby earning foreign currency. The Chongqing plant personnel said that it would be best if Cummins were to agree to equity investment in their factory.

Chongqing managers said that they had no plans to import foreign diesel engines from additional sources. However, MMBI officials said that China had licensed technology for the manufacture of trucks from an Austrian company, Steyr-Daimler-Puch AG, and this agreement with Steyr included the diesel engines that powered the trucks. The production volume, actual or anticipated, could not be ascertained.

In the future, China aimed at a balance between gasoline and diesel engines to power its vehicles. In the past, production of gasoline engines had predominated and therefore the manufacture of diesels was to be expanded. However, diesel fuel was in short supply because the price of gasoline was much higher than that of diesel fuel, and this circumstance discouraged the production of diesel fuel by the country's oil refineries. In order to overcome this difficulty, pricing policies and economic aims had to be better coordinated, according to MMBI officials.

As described, the Chongqing plant had organized its own maintenance and service organization, independent of the Technical Service Center established in Beijing. However, Chongqing managers said that they hoped to establish cooperation with this center in the future and to contribute to the expansion of its activities.

CASE 7

Combustion Engineering: The US. Perspective

INTRODUCTION

In 1986 Combustion Engineering (C-E) was established as one of the pioneers in transferring technology to the power-generation industry in China. Entered into by the Power Systems Group of Combustion Engineering and the Ministry of Machine Building Industry (MMBI) of the People's Republic of China (PRC), a 15-year licensing agreement for the dynamic transfer of boiler technology was signed in Beijing on November 21, 1980. Under this agreement, a 300-megawatt controlled-circulation boiler unit was to be co-designed and built with Shanghai Boiler Works and a 600-megawatt unit with Harbin Boiler Works. The acquisition of boiler technology was part of a broader power-generation program wherein China was also acquiring compatible generator technology from Westinghouse, USA, and assistance from Ebasco, USA, as general contractors in installation.

This case describes C-E's experience in licensing technology to China and the process by which C-E managed the negotiations, agreements, start-up, and ongoing transfer of technology. (Exhibit 1 summarizes the major events in C-E's history in doing business with China.) The case is based on a series of interviews with senior executives of Combustion Engineering and its Power Systems Division.

Combustion Engineering

The origins of C-E go back to 1914 when a firm by that name was founded in New York to manufacture fossil-fuel boilers for the utility industry. In its early years, C-E made a technological breakthrough that enabled coal to be pulverized to the consistency of talcum powder. This powder could then be used in stationary boilers, thus greatly increasing the combustion rates and efficiency of the fuel-burning process. C-E also developed a new furnace utilizing walls made of tubes through which cooling water circulated. Existing refractory materials and furnace designs could not withstand the higher temperatures of the pulverized coal. These two developments in combustion technology provided the base for C-E's initial growth in the manufacture of furnaces and fuel-burning equipment in the 1920s.

After several years of intermittent struggle and moderate commercial success, C-E found its normal business activity curtailed by the onset of World War II. During the war, C-E produced marine boilers for ships, ultimately supplying about 5,000 steam-generating units to the U.S. Maritime Commission, which was roughly half the total supplied by all manufacturers. C-E also built power boilers for the giant atomic production plants at Oak Ridge, Tennessee, and Hanford, Washington.

After the war, as part of the Marshall Plan, C-E built and installed 65 major steam generators in 15 war-torn countries. Nuclear energy had also emerged as a result of World War II, and C-E was quick to take up this new challenge as another facet of its expanding power-production business.

After World War II, decimated European industry turned to the United States for help, resulting in a sharp increase in C-E's direct export sales. However, as Europe rebuilt its own manufacturing facilities, direct exports declined. To take advantage of this changing trend, C-E took steps to export its advanced technology rather than its hard goods. European technology had stagnated during the war, and they required engineering and design assistance.

Accordingly, C-E built up its licensee organization, licensing independent companies in various countries to manufacture steam-generating equipment in their own plants using C-E designs. The first of these agreements was signed with Franco Tosi, S. A., of Italy in 1949. Within five years, C-E had license agreements with companies in England, France, Germany, Holland, Sweden, Norway, Italy, Argentina, and Japan. This network, plus direct sales, put the company in a strong position the world over. By the start of the 1970s, C-E executives estimated that 40 percent of the Free World's steam was generated by equipment of C-E design.

There was, however, another reason behind C-E's decision to concentrate on licensing overseas companies. Heavy industry requires a huge investment to set up a plant and production facilities, and it is seldom practical to start a new company overseas. The C-E philosophy, therefore, was to look for the best firm with heavy equipment capabilities in each country, license it to acquire C-E technology, and provide it technical and management assistance.

C-E's sales volume in 1963 was $357 million, increasing to more than $1 billion by 1971. During this period, C-E's sales of steam-generating equipment to electrical utilities doubled. However, as a result of C-E's corporate diversification program, these sales accounted for only about 34 percent of C-E's 1971 sales volume, as compared with nearly 50 percent before 1963.

By the end of 1985, C-E was a diverse engineering company with 32 different licensees throughout the world. Its total annual company sales totaled $2.41 billion. The company employed 24,761 people and operated 30 domestic manufacturing and processing plants. Sales revenue came primarily from three broad categories: (a) equipment, products, and services supplied to industrial markets; (b) design, manufacture, installation, and servicing of steam-generating systems and equipment for the electric utility industry, including nuclear fuel steam supply systems; and (c) design, engineering, and construction services, primarily for the chemical, petrochemical, and petroleum industries. The majority of C-E revenues, 87 percent of the total, came from domestic sales. Of the balance (13 percent), 6 percent came from Canada, 6 percent from Europe, and 1 percent from other foreign countries, including China. The foreign revenues consisted mostly of royalties or license fees from C-E's licensees in various countries.

C-E had evolved into a global presence by constantly developing worldwide markets based on technology and market familiarity. Frank Marshall, vice president of Licensing and Joint Ventures within the International Division, cited the reasons why C-E put so much emphasis on licensing. These included the following:

- □ Capital requirements are very high for the design, engineering, and construction of power plants.
- □ Since power generation is a core industry, most countries desire to maintain control over their power-generation or boiler facilities.
- □ Licensing leads to other related business over the long term such as parts sales and service.
- □ Royalty flows as a percentage of sales price can be quite substantial over a minimum 15-year or longer contract period.

Joseph Condon, corporate vice president of C-E's International Operations, added a different perspective to the use of licensing as a long-term business strategy:

For us, our technology is really everything we have to sell. That really is our lifeblood and soul, and we have to protect that technology in the long term. Not only do you practically give it away to countries like India, but it comes back to haunt you because if you have a really good licensee, and the only kind you want are those that learn quickly and become market driven and are aggressive, they are not satisfied with staying in India or staying in China; they want to then come into world markets. So you have them coming around to bite you in your own other markets.

C-E's Licensing Organization

C-E established a specific licensing department dedicated to servicing the family of licensees. Within this department, certain individuals were designated as sponsors. The main task of a sponsor was to handle all of the details and servicing related to the contractual obligations of three to four licensees. Ongoing customer service was handled through these sponsors. Alex Sivas, vice president of Business Development and Sales, International Division, explained:

> All correspondence comes to the sponsor. Let's say the Shanghai Boiler Works wants to send someone for specific training. They write to the sponsor and say they want to send somebody for training. The sponsor makes the arrangements, sends them the agenda, and that's it.

In some areas of the world, such as in China, C-E assigned a sponsor for just one licensee. The sponsor monitored what the licensee received and ensured that information was received on time and in the form requested. All of C-E's affiliations were dynamic licenses so that the sponsor also supplied information to the licensees periodically on new developments in their particular field. As Alex Sivas stated:

> We have had licenses since 1950 that have been renewed three times after expiration. Mitsubishi Heavy Industries, today, is one of the biggest producers of steam generators in the world, yet they have maintained their licenses with us, and they renew them. The reason they renew is to take advantage of C-E's R&D. We also have joint R&D projects that we do together. There are also certain areas of expertise they do not have and must use our name in order to do business. For example, during the 1960s and 1970s, the Japanese converted from coal-fired units to oil-fired units. Their expertise was concentrated on oil firing. All of a sudden, the 1973 oil embargo came along, so everybody had to convert to coal. In the meantime, we had continued to design both types of units, coal and oil. We had vast experience with coal firing and the Japanese did not. If they wanted to do any business overseas, they had to prove they were tied up with a company that knew what they were doing in the coal-firing end of the business.
>
> The result, however, is that there are problems with international competitiveness that cannot be covered in an agreement. We cannot even prevent them from coming to the United States and bidding against us. We tried to restrict the Japanese from passing information over to communist countries, but were not able to restrict them. In essence, our business concept is to be wide open, the fountain of expertise, to stay ahead of the technological game. We are not really ahead of our licensees in terms of technology because we are continually updating them with our current state-of-the-art technology. They fall behind only if they do not update their licensing agreement. That is our philosophy.

DEVELOPMENT

Early Contacts

Combustion Engineering's first introduction to China took place in 1938, through an affiliated engineering and construction firm by the name of Lummus in which C-E had part ownership (subsequently increased to 100 percent). Lummus was one of the major design engineering firms in the world and was responsible for the complete design, engineering, and construction of an oil refinery in China. This successful project subsequently led to three additional projects for Lummus, including the first modern ethylene plant built outside Beijing.

From those early successes, C-E established in the minds of Chinese engineers its capability as a supplier of design, engineering, and construction services. C-E also realized during these early years the strategic importance of China because of the huge market potential that existed for power plants. This business could not, however, be fully developed until U.S.-China diplomatic and trade relations improved.

Alex Sivas recalled the activities that led to C.E.'s technology transfer agreement:

> Our real interest in entering China as a strategic move goes back to 1976. We had heard that some of our European licensees had gone into China and had been successful in selling them power plants. These are Western European licensees, so we knew that the Chinese were beginning to open up. We wrote letters to several ministries and institutions, but never got an answer. We also tried to use intermediaries who were going in and out of China, but even that did not work. There was complete silence. Finally, one of our Chinese employees here at C-E went to Shanghai to visit his relatives. He started talking and expounding the virtues of our technology and C-E on his own, with no authorization from the company as such.
>
> After that, in 1977 we received a letter from the Ministry of Machine Building Industry (MMBI) stating that they were interested in hearing what we had to say. We formulated a fossil and nuclear seminar, and in November 1978 went to Beijing and talked to the MMBI people. Dr. Herb Cahn took care of the nuclear portion of the seminar, and Sam Blackburn took care of the fossil portion. We gave them presentations. Of course there were other people there from the State Planning Commission, State Economic Commission, Ministry of Electric Power, and some of the other design institutes. This two-week seminar was very well attended.
>
> In December 1978, the United States normalized relations with China, and shortly thereafter the Ministry of Machine Building Industries (MMBI) sent two delegations to visit us. We spent a lot of time, maybe two to three weeks each time, discussing our technology and manufacturing capabilities and our market share in the United States and abroad; they really took a lot of information from us. We finally got an invitation to start negotiating a technology transfer agreement.

Sivas described the early meetings with the Chinese negotiators as very one-sided:

> C-E was always giving out information and rarely getting any in return. The Chinese negotiators were very vague about their requirements, the capabilities of their shops, what they were manufacturing, and what designs they were currently using.

Chinese Visit to Combustion Engineering

The first delegation that came to the United States from the MMBI consisted of twelve people including engineers, accountants, and officials. The technology transfer negotiations began in February 1979 and were concluded almost 22 months later, on November 21, 1980. During this period, Alex Sivas, the key individual responsible for the consummation of the agreement, made more than 20 trips into China, usually for a minimum of one week to a maximum of seven weeks.

According to the November 1980 agreement, C-E agreed to supply MMBI with two controlled-circulation steam generators designated "verification units." A 600-megawatt unit was to be built in Harbin, and the other, a 300-megawatt unit, was to be built in Shanghai. Although C-E was known worldwide for its premier technology, China wanted to have these verification units built on-site in China as a means of testing C-E's technology. In addition, building the units in China would allow the Chinese engineers and technicians to learn firsthand the design, engineering, construction, and management techniques used by a U.S. corporation.

Under the contract, C-E was responsible for the design of the two units, and 30 percent of the components to be used in those units. Seventy percent of the components were to be provided by local Chinese plants. Eight hundred man-months of training and 70 man-months of technical assistance were also included in the 15-year agreement.

NEGOTIATIONS

Venue and Personnel Involvement

The initial contract negotiations took place exclusively in China. From the Chinese team's standpoint, this was the most efficient because they were required to report progress every night to the MMBI and receive appropriate instructions. For C-E, this arrangement was not as convenient. Bob Haun, who worked as the China Project manager in the International Division of Power Systems after negotiations commenced, and who had 20 years experience in the Nuclear Division of the company, described the scene of the initial negotiations:

It was a very formal setup. The meetings were held at Erligou (two-mile gully), which was a common place for these types of meetings. There would be an interpreter and about 50 Chinese in the room. A good number of the people at these meetings eventually came over here as engineers. Their interpreter was just that, an interpreter. He wasn't an engineer and had no technical background. The interpreter was very good and was involved in all of the negotiations for the contract from the beginning.

From the beginning of negotiations, there was a basic "club" of 12 people from C-E who were involved and who traveled back and forth from the United States to China. Two to four C-E personnel were generally present at one time, with the actual negotiation personnel rotating periodically. There was always a lawyer, one or two engineers who were specialists in either boiler or control manufacturing, and a person from the licensing engineering office. C-E always brought their own interpreter to ensure their ideas and points were clearly stated. Sam Blackburn, who was working as director of Licensing Engineering for the Power Systems Group, described the reason for rotation:

> We tried to have R&R for the negotiating team. I don't mean this in a very negative way, but Beijing is not my idea of Paris in the Orient. After three weeks to two months, in some cases for some people, you needed to get out. The contract negotiations were ongoing, so somebody had to be there. We had kind of a rolling membership in this little club, if you will. It never got past a certain number of people, but there was always some interchange.

The interpreter used by C-E was an ethnic Chinese engineer who had been with the company for many years; he was the initial contact in China responsible for getting the Ministry interested in C-E technology. His wife was a professor at the University in Stamford, Connecticut, who taught at the University of Beijing during the summer. He was not just an interpreter, he was also a cultural monitor. Alex Sivas had the following to say about the role of C-E's interpreter:

> He helped us with the hotels, the food, and so forth. But he did not tell us what was going on behind the scenes. I wanted him to come in at night after we had finished our whole day's discussion, and I wanted to find out what they were talking about, what is the atmosphere, are things going well, or are they going bad. He would never tell me. I used to get so mad because I did not understand the language, and my only means of knowing what they were saying was through this guy who would never tell me what was happening. We could have profited from someone who was more of a monitor, who was just sitting there observing and watching the dynamics of the situation, so we did not feel so isolated.
>
> We also used consultants before we got involved in China, and there were even some Americans who could speak Chinese. Perhaps they could have been useful in this situation. Chase Bank wanted to represent us, but we had already established our contacts and knew what we were doing and where we were going.

Negotiating Styles

Bob Haun, who was involved throughout the negotiating process, recalled that his first trip to China to negotiate the terms of the technology transfer was not successful, and it looked as if an agreement would never be concluded. The Chinese initially were difficult to deal with, in part because of the language barrier. It was thought that this was only a negotiating trick. In Haun's words, "They would say, 'I don't understand what you mean,' when, in his opinion, they knew exactly what you were saying."

Bob Haun summed up his frustrating initial experience dealing with the Chinese negotiators as follows:

> I'm sort of a get-the-job-done type of guy, and I am very American. I did not appreciate the slowness of the Chinese. Every meeting started off with, "Give us what you want for the agenda for the meeting," and they would get that, and they would go over it and over it, spending a whole day and then would come in the next morning and say, "We went over your agenda, here is our agenda, now let's go over our agenda," and I'd look at it and say, "Ok, that looks good, let's use that agenda." They would say, "But it's different than your agenda." [I'd answer them,] "Well, it is just in a different order, but it's basically the same thing." They would say, "Well, you had better go study that." It was difficult, and I realized that we had a delivery schedule. This was not a job that was open ended. We had to get equipment out by a certain date, and we had only so long to get the unit designed.

Confrontation during negotiations was one area that C-E attempted to avoid during the two-year process. On one occasion, Alex Sivas felt that progress was much too slow and chose to take an action that would accelerate the Chinese decision process and still save face for everyone involved. He recounted the following incident:

> I was acquainted with the commissar, who was also a Party member. I called him one day and asked him if he would come to the CAAC terminal at the airport to assist me in making reservations to leave the next day. I did not announce it at the meeting because I did not want to embarrass the chief Chinese negotiator across the table. Within half an hour after I made reservations to leave, the chief negotiator was on the phone. He said, "Are you leaving Beijing so soon?" I said, "Yes, something came up and I have to get back to the States." This was toward the end of the negotiation. I left, and about a week later they called me back and we settled the deal. You cannot let them lose face, and you cannot put them on the spot by saying that you either agree with me or I am leaving. It does not leave them any way to move. I knew that guy was going to report immediately what I was doing.

Officials from MMBI would often invite C-E officials to banquets and would pontificate at length. According to Alex Sivas, the favorite expression of the head of the Chinese team was, "You must open the window and let some fresh air come into the negotiations." In other words, cut the price.

When negotiating price, the Chinese side would never give an indication of the reasonableness of C-E's offer. As Alex Sivas explained:

> If C-E wanted $20 million and the Chinese only wanted to pay $15 million, they would not say you are too high at $20, you have to bring your price down to $15. They would say, "You are very expensive. Your competitors are more reasonable people." It was maddening because they would never tell you where you stood. They would go to our Chinese interpreter, the guy we brought from here, and would pass him a little information on the side in Chinese, and he would come and tell me that he heard this and that. They don't have more than $20 million in the till and you are asking $30. The information came in a very circuitous way, never head-on. You have to keep your ears open to know what is being passed on to you.
>
> The Chinese calibrated their price by playing different competitors off one another and off C-E. They had Babcock next door, and when they were negotiating with Westinghouse, they had G.E. in the wings, so it was a madhouse. They had us at Erligou, and we were sitting in different rooms, and it was a zoo. They would come back and tell us if Babcock lowered its price.

C-E did manage to negotiate a slight premium over Babcock and other competitors because it was the engineering leader at the time. That premium, however, was very small.

There were stories that the Chinese side occasionally bugged the phones of their negotiating counterparts. These rumors were never substantiated; however, precautionary measures were taken by C-E personnel who would fly to Tokyo if there was anything of a very confidential nature that needed to be discussed with the home office.

The Chinese were considered very good negotiators; they were very patient, very slow, very methodical, and very detailed. Most of the Chinese negotiators turned out to be people who would ultimately be involved in the various phases of the design, engineering, and construction. A representative from the MMBI was their chief negotiator, and the team always included a representative from the foreign trade office.

Choice of Product and Technology

C-E's general policy was to offer and make available state-of-the-art technology to its licensees, regardless of the level of development in the licensee's country. Sam Blackburn's role during the negotiations was to ensure that the right type of technology and hardware was selected to be transferred to China. Blackburn summarized C-E's policy as follows:

> If some licensee companies do not have either the manpower or the money to spend for the training, obviously they are not going to be able to utilize the most sophisticated technology. However, it is available to them if they want it. The

stuff that goes out, our paper, is up to date, but whether they can use it is a function of how much time they are willing to spend in training.

The Chinese training program we agreed upon was about 800 man-months. When we licensed BHEL in India, our second most extensive licensee to China, there was something over 300 man-months of training involved. This Chinese training program was massive, in part because they had so much more to catch up with.

In addition, they didn't care if it was useful to them or not, if something was part of the technology transfer they wanted it, even if they were not going to use it. That to me was very frustrating. As an engineer, I try to be practical. That was the most difficult thing I had. If you don't really see a need for that, why do you want it? "Just because I want it" was their response.

The C-E agreement involved a patented design called "controlled-circulation steam generators," which could be adapted to a variety of fuels and came in a variety of sizes. The Chinese opted to take the 300-and 600-megawatt power plant designs as their standards. Alex Sivas made the following observation:

> Even though they knew that we were the premier steam-generator company in that range of equipment, they still wanted to prove the design, so they gave us two orders for a 300- and 600-megawatt unit, which they called the verification units. What they wanted to do was verify that our technology was in fact operable. They already have two units from our licensee in Italy, Franco Tosi, which are of the same design, and two 350-megawatt units from Mitsubishi Industries, which are operating satisfactorily, but they still insisted on getting verification units.

The 600-megawatt verification unit was to be produced in Harbin, and the 300-megawatt unit in Shanghai. These sites were chosen because of considerations of size of production facilities and availability of experienced technical personnel.

C-E was to co-design the units with Chinese Boiler Works (Harbin and Shanghai) engineers. The Chinese side considered co-designing most beneficial because it was considered the most effective process for transfer of technology. The co-designing was possible because each boiler needed to be specially designed to suit local fuel characteristics. During the production of the verification units, the Chinese engineers were also to develop and learn new design and manufacturing know-how; welding techniques; manufacturing management; and metallurgy, including stainless and chrome alloy types of carbon steel. Further, they were to "verify" the feasibility of the manufacture of the C-E product under local conditions. The Chinese engineers were considered to be excellent but had not been exposed to current Western technology.

Renegotiation and Conclusion of Agreements

C-E's negotiators had concluded what they thought was an agreement with the Chinese side well before November 1980 and had submitted it for approval to the Chinese State Economic Commission. To their surprise, C-E's team was informed that the Chinese negotiators had exceeded their authority and had not used the correct format for the agreement. The result was that a new Chinese negotiating team was assigned in midstream.

Documents already agreed upon and signed were discarded. The new pesonnel who were added to the original Chinese negotiating team were basically commercial personnel. The original negotiating team members, who were engineers, also continued to be involved. The new agreement negotiated with the enlarged group was much more specific and detailed, compared with the agreement that had been tentatively concluded with the original negotiating team. In fact, the documents, instead of being one inch thick as in the first agreement, became closer to three inches thick. As Sam Blackburn recalled,

> In most license agreements, we can generally make statements such as "information shall consist of drawings, files, etc. If it is readily available, you are entitled to it by asking." The Chinese license is interesting because it contains language like, "shall include, but not be limited to" and then also specify an exhaustive list. Previously in the agreement, we may have already mentioned they will get "all drawings," but they still want this extensive list.

When asked what finally clinched the deal, C-E officials indicated they had sensed from the beginning that the Chinese wanted C-E's technology and design assistance. Apparently the MMBI was also interested in General Electric (G.E.), but saw G.E. as very inflexible with its price. G.E.'s initial approach, according to some executives, was to tell the Chinese to "either take its offer or leave it." C-E executives observed that, in addition to a good product and company reputation, flexibility in terms was ultimately the key factor that led to C-E's successful conclusion of the agreement.

When asked if he would do anything different in negotiating with China, Alex Sivas responded as follows:

> Not really. I don't know how differently I could do it. You really can't plan ahead, you just have to meet the negotiation head-on as it progresses. I don't know what I could do to prepare myself. The Chinese are masters at it. You have to have tremendous patience.

On November 21, 1980, C-E and MMBI signed the general agreement. Sam Blackburn summarized the outcome by saying that this agreement was completely different from any other previous agreement C-E has made with another country. The C-E agreement consisted of four main elements: (1) a definition of the technology to be transferred and payment terms; (2) the training of the Chinese technical and management personnel in the United States; (3) the boiler and power systems components that C-E would supply; and (4) the provisions for updating the agreement and adding other technology, products, and services.

Payment Arrangements

Among the biggest hurdles of the negotiations were the price and payment terms. At the time the contract was signed, China did not have a serious foreign exchange problem; the Chinese government had enough cash on hand to pay the entire purchase price. The license agreements for C-E and Westinghouse were both financed, however, by the Export-Import bank for $28 million each. These transactions were the first financing schemes by the U.S. Export-Import bank in China. The interest rate charged was 7.6 percent when the going rate was 8 percent. The Chinese officials commented that 3 percent would have been a more reasonable rate. The Chinese government wanted to set a precedent for being able to demand credit from the bank whenever it desired. In years past, the Chinese tried to stay away from banks due to the high cost. An industry combine supported by the Italian government loaned money to China in the past at 2.5 percent.

The contract called for specific activities that were to take place before any payments were made, for example, target dates for shipping of document packages were specified. As shipments were made and the contents accepted, payments were to be made to C-E.

START-UP

Start-Up Personnel

Sam Blackburn, who was involved from the start of negotiations, became the project leader to monitor the overall flow of technology and to make sure deadlines for the different transfer stages were met and no late penalties were incurred. Reporting to Sam Blackburn were Anthony Czapracki, who was a consulting engineer in international licensing, and was responsible for the implementation of the licensing and technology transfer portion of the contract (software), and Bob Haun, who was in charge of implementation of the components portion of the contract (hardware). There was some overlap of responsibilities, however, due to engineering work that needed to be done for components under the technology transfer agreement. Essentially Sam Blackburn's role was to maintain a quality check to make sure decisions made by Czapracki and Haun on the one hand, and the Chinese side on the other, were consistent, timely, and effective. Sam Blackburn commented that his role did not fit clearly into a job description because it was such an unusual job.

During the early stages of organizing the project, two other individuals were dedicated to working on the Chinese license agreement, one person from manufacturing services and another from licensing engineering. Their jobs entailed planning and implementing the first phase of the contract, transferring a great deal of paper such as standards, manuals, and drawings for various designs.

C-E agreed to do a small portion of the engineering work for the first two verification units; however, most of the work was to be done by the Chinese personnel under C-E's supervision, according to the co-designing agreement. The detailing for the jobs in these first two projects was done exclusively in China, since the Chinese had had experience in boiler design and construction.

Export Licensing Considerations

Anthony Czapracki described the initial approvals that had to be obtained from the Chinese and U.S. governments before the actual transfer of technology could take place:

> One of the first things we had to do was satisfy the Chinese side that the requirements of both U.S. and PRC governments would be met and approvals obtained. They did not tell us what their government approvals were. On the U.S. side, since China was listed under a special classification, everything had to be put through the Department of Commerce, State Department, and the Department of Defense. I spent the first two months between here (Windsor, Connecticut) and Washington, D.C., and the Chinese, at this stage, were spending a lot of time in the engineering offices of Westinghouse in Lester, Pennsylvania. I traveled between these three locations trying to get a consensus of what constituted an approval in their eyes trying to convince the Department of Commerce that what we were exporting was not high-tech exotic, but at the same time convincing the Chinese that this was the best technology available.
>
> Because of the tight schedule that the Chinese had insisted upon for themselves for the first start-up of the verification units, they put themselves on a very tight time frame which required a bit of coordination between C-E, Westinghouse, and Ebasco, the acting general contractor. Because the payments and other arrangements hinged on the effective date of the contract, it was important to get the necessary approvals as soon as possible.

Since work could not begin until the export license was issued by the U.S. Department of Commerce, and because the approval was taking longer than expected, Anthony Czapracki spent considerable time trying to renegotiate a start-up date with the Chinese side. According to the contract, if the approvals were not in place within 90 days of signing the contract, the contract would be null and void. Prior to obtaining these approvals, Czapracki recalled the interaction he had had with the Chinese and the extremely cautious behavior they exhibited:

> I knew I was getting close to the date and was not getting very much positive feedback from Washington as to when they might give the approval. I was trying to get into a position of saying, "Gentlemen, the contract as written, the language you wanted in here, says that after 90 days, it's null and void. I don't want to see that happen. I want an extension. If I don't get the approval by this date, I want to be able to sign this extension to give us more time for the approvals."

At the same time I was worrying about government approvals from the Chinese side and how long that would take. They seemed to indicate no worry.

The gentlemen I was dealing with gave us every indication that they had signing authority. These were the same two people who had signed the license agreement with Alex Sivas. Li Peizhang from MMBI was the chief negotiator. We were down to the last Friday before the expiration date when I met the group in Buffalo. They were reluctant to sign anything.

We were in the "early days." They were very meticulous about their wording and very suspicious about what any implications meant, especially since we were dealing in English. I met them in Buffalo on a Thursday, and they finally signed the extension agreement, which turned out to be unnecessary, because as soon as I got back to my office, there was a message waiting for me that the Washington Office had received the approvals. The Chinese were basically pushing us to get some kind of indication of ability to export to China. They had no problems on their side, in retrospect. They had the allocation of money and were able to keep their dates with payments.

They were returning to China shortly thereafter and wanted us to sign a Chinese version of the license agreement, although it was dictated to be in English, and the English copy would prevail in arbitration. They insisted for their use, that we also sign a Chinese copy of the contract. We were a bit reluctant because we had no part in putting it together or translating it, or no time to review the whole document, so we ended up reviewing key documents and signing it. I delivered that to them in New York as they were getting on the plane. It went down to the last hour. They returned to China with the assurance that we had the government approvals and the signed copies and that we were proceeding with the documentation and transfer of the paperwork.

Dissemination of Documents

The months of March, April, and May 1981 were spent transferring documentation to China. Initially, C-E had some difficulty with the details required by the Chinese side. All documents were required to be addressed in Chinese characters. Once they were dispatched, the Chinese side needed to know when the documents would arrive, on what airplane, and what air bill number, in order to pick them up. Packages had a way of getting lost or neglected very easily in China unless someone knew exactly what to look for. C-E was put in a position of having to worry about every detail along the way until confirmation of arrival was received.

As Czapracki recalled:

There was a reluctance for them to say, yes we did get the box, because they did not want it to imply that yes, everything in this batch is what is in the contract. There was a great deal of suspicion on their part in those early days, in those early batches. It resulted in a lot of frustration.

It took a full-blown effort to meet the dates and the volume of documents they requested. They seemed to place an importance on a vast quantity of material. In talking with them, I said, "Gentlemen, what do you need first to get started?"

Typically that would be the manufacturing information specifically for the units they are going to build. But they wanted everything up front whether they needed it or not.

I was in Harbin that same summer and heard of their experience with Russia in the early 1950s and 1960s, when projects were left half finished and all of the drawings were taken back by the Russians with no copies left behind. I can see why they have that suspicion. We tried to be as open as possible to satisfy them in the beginning days.

We also tried to remind them that they had a rather tight schedule set by themselves and that they really should be concentrating their efforts in certain areas. They proceeded to do it in their way and in their own style.

All documents were addressed to Li Peizhang of the MMBI in Beijing. Every paper document sent was required to have four copies. Rather than have C-E send documents directly to the appropriate Boiler Works, MMBI requested that everything be sent to one place. They wanted to count and inspect every shipment to make sure it was all there and then handle the distribution themselves.

There was a provision in the contract that stated that duplicates could be requested if anything was lost, unclear, or could not be read. C-E made it a policy to fulfill their requests, if prior shipments were claimed to have been lost or misplaced, whether it could be proved or not. The following comment was made with regard to paperwork:

> We have saturated them with so much paperwork that they do not know what to do with it. I have been there and seen it stacked up in rooms and libraries.
>
> A lot of it was just confidence building. The relationship now is much better. We have gotten on with other projects, and there is a better sophistication on how they are going about accepting the technology transfer.

By 1982, C-E had made all their shipments on schedule and were starting to pick up on lost and unclear documents. They continued to make shipments of new information and even some revisions of material sent only the previous year. Project managers involved in the document transfer commented that they were certain the Chinese opened up information on revisions before they had opened up the first version. As new standards were issued by C-E, they were immediately disseminated internally to all licensees. Mail delays not withstanding, all licensees were very current. C-E maintained a worldwide policy of sending whatever documentation a licensee needed, even if it had already been sent earlier. Even though it was inefficient at times, C-E policy stated that, if it solved the licensee's problem, it should be done.

Concluding his thoughts on documentation, Sam Blackburn made the following observation:

> We have a general set of standards that go out to all licensees that are rather massive. We are a paper factory. On top of that, we have an almost infinite number

of parts. The Chinese are the only ones who have ever received every drawing we have available in the entire company within their product line. We were sending tons of drawings to them, where we would only send pounds to other people.

Training Programs

Training was a very important and integral part of the technology transfer package, without which the documents had no significant meaning. Following the signing of the contract and prior to the arrival of the first group of Chinese trainees in the United States, some C-E personnel feared that the Chinese would cancel the contract after a lot of work had already been done by C-E. This fear became stronger when the first wave of 12 trainees failed to arrive in June 1982 as originally agreed upon with Li Peizhang. Czapracki recalled,

> They [the Chinese] kept assuring us that canceling the contract is one thing that would not happen, that energy was very high on their priority list, and that their money had been set aside. Again, we got that message, but no signs in writing that could make us feel a whole lot better. We were going by trust along both sides. We were always looking over our shoulder thinking, what happens if they notify us or if they don't notify us and here we are still continuing with the stuff in the pipeline. So we went ahead on good faith, and the first nine engineer trainees eventually came over, three months late, in September 1982.
>
> The design engineers started in our Windsor, Connecticut, facility and were in the United States for a total of 18 months. In the meantime, we also had 50 to 60 other people scattered all over our shops in Chattanooga, Tennessee; St. Louis, Missouri; and Chicago, Illinois, learning the fabrication techniques of different parts of the boiler. We offered special programs for each group in order to help them learn the whole gamut of our fabrication techniques.

The first problem related to the training consisted of having to convince the MMBI of the project sequence and of the number of people to send over. C-E suggested that it was most logical to begin with training of manufacturing, but the Chinese insisted that R&D and design engineering training be undertaken first. The Chinese got their way. Czapracki made the following observation:

> It typically gets down to negotiating: "Gentlemen, this is what I have to offer this year." They tell me what they want, and we would go from there. They would insist that every January or so, they had to have a plan for the next year's training. That took a good portion of my time preparing to negotiate a final agreement we could live with. We would usually sign a protocol or sometimes an amendment to the basic agreement with a schedule. Without a doubt, even sometimes before the ink would dry, in good Chinese tradition, as I have heard other people say, they would ask for changes. We had yearly protocols. There was a provision in the technology transfer agreement that our top management (the corporate chief executive officer, the corporate vice president of Engineering, and corporate vice president of International) would meet on an annual basis, in the

United States one year and in China the next year, alternately. The Chinese would usually send someone of the level of Li Peizhang. That was very important in the early days for them. We still maintain that tradition, but we don't get nearly as excited if it gets canceled and neither do they.

The Chinese always insisted upon equal representation or equal access by both boiler works—Shanghai and Harbin. Even though it may have been more convenient or more timely to get some additional training for the Shanghai people because of their one-year earlier start-up schedule, it had to be done on a parity basis. In total, the contract called for 800 man-months of training of the Chinese personnel for both verification units.

Two of the first trainees were involved in computer engineering. Their task was to learn about C-E's computer programs. C-E predicted marginal results given the short time they had allocated for this learning and their inexperience with modern computer technology. These individuals came to C-E expecting to learn how to input the program, do the calculations, trouble-shoot the program, and retrieve the output. They intended to stay nine months and ended up staying a full year and still did not cover everything.

On-the-job roles of these Chinese trainees were very different as each person focused on learning a different type of expertise within engineering. Except for the two men in computing, two trainees were teamed up with one C-E training supervisor. This was useful, especially when one person's English was not as good as the other's, and it also allowed C-E to train two people at once. One of the trainees was usually designated as the interpreter. C-E did very little classroom training.

The Chinese side was very concerned about the living costs of the trainees in the United States. Visas were generally issued for a maximum period of three months. Obtaining visa extensions took a great deal of time and effort on the part of C-E staff. Because the licensees were not familiar with Western ways, the Licensing and Joint Ventures Department provided quite a bit of service they normally would not provide. Accommodations had to be found, and transportation had to be arranged because they were not allowed to drive cars. Many other social graces had to be extended that were not normally required in servicing licensees.

After the Chinese trainees had been in the United States for some weeks, they were told they needed to get a driver's license and to provide their own transportation. A couple of them did take driving lessons, which was very progressive on their part. When their local representative came to the United States from China, however, he sent for a driver from China, since he did not want the engineers to drive. Bob Haun recalls the conversation he had with the Chinese representative:

> I asked him if he had made arrangements for transportation, and he told me they were just going to get bicycles. I said, "Well, I don't think that is going to work. It would be very healthy to do it in the summer, but two things come to mind.

First, in the winter, you're never going to get around, and second, this road is very busy, especially in the morning. There would be accidents and people could get hurt." I guess they went back and talked it over in one of their meetings which they had daily to review everything they did during the day. This one guy seemed to control the purse strings, but he was not one of the head guys. They did eventually get a van and a driver. The guys who learned to drive did buy a car. The first place they went to visit was to West Point.

Bob Haun recalled the age differences of the people that came to the United States for training.

In the very early days, I don't think I saw very many people under 48 or 50. I remember thinking of myself as quite young, not sporting any grey hair. That was a disadvantage in trying to be a teacher to these people. I had a credibility gap. As we have gone through 1981, 1982, 1983, and 1984, there have been some younger people. Not only need, but seniority influenced who was placed first in line for the American training programs. Later, we saw the younger, less senior people coming through. We did extend some training to the ultimate customer, the Ministry of Water Resources and Electrical Power (MWREP), under the 800 man-months. They had 44 engineers in all during 1984 and had by far the widest age distribution. I think the youngest person from this group was 21 or 22. He was a graduate engineer trainee for boiler operations.

All of the engineer trainees were married, but their wives stayed back in China. The group was very close-knit, and there were no defections from the C-E group of trainees.

The Chinese trainees were anxious to get their full money's worth of supplies here in the United States. When they arrived, C-E supplied them with the textbooks and other miscellaneous supplies they needed. The Chinese leader requested that C-E supply the trainees with calculators and little desk computers. Bob Haun refused their request. He indicated that C-E did not supply any of their engineers with those items. They were available for use, but were not given out personally because they could be stolen. The Chinese response was that Westinghouse had given them a box of 100 calculators. Bob Haun replied,

"How many people do you have at Westinghouse?" He said, "40." I then said, "Well, send 60 of those calculators up here." The Chinese were a little pushy, in part due to the $30 million they were paying us. They wanted something for their money. That is really not a lot of money, even though it sounds like a lot. That sum included everything for the whole 15-year period. Part of the problem is that you are talking with a person who is making $15 a month in salary. I think I had some appreciation for that.

In 1985, the Chinese notified C-E that they would not be sending any more people to the United States for training. They needed them in their shops in order to complete the detailed drawings and to put out the product they were building. By this time, they had used up about 400 man-months of their training allocation.

Equipment and Component Purchases

Under the terms of the contract, C-E was responsible for the conceptual design of the boilers together with the Chinese side. About 32 Chinese engineers in Windsor, Connecticut, participated in the design phase. Co-designing was a critical part of the transfer process, since C-E transferred technology through co-designing the verification units. C-E also agreed to furnish about 30 percent of the components, and the Chinese were to finalize the detailing of the design and build 70 percent of the components. There was a lot of give-and-take regarding what C-E thought the Chinese plants could successfully produce in China, especially after C-E people saw their manufacturing facilities. The specific components that would be supplied by each party were spelled out clearly in the components contract.

After the Chinese engineers started doing the detailing of the boilers in China, they ran into problems. Their belief was that C-E would give them the manufacturing drawings necessary to make all the parts in China, but found out that this was not part of the deal because many of the parts C-E used were purchased from other suppliers. As a result of this misunderstanding, additional negotiations were required. There were a number of items the Boiler Works just could not source in China, such as solenoid valves, switches, piping tees, and many special fittings for duct work. C-E had to send specialized steel squad supervisors over there to design the steel with them. The heavy sections of steel found in the United States, such as wide-flange columns made by U.S. Steel, were not available in China. Since they did not want to order steel from abroad, they built up shapes in a rather intricate design to overcome this deficiency.

As time passed and the project progressed, the Chinese determined that more and more parts had to be purchased through C-E. Separate negotiations for these components were made on an as-needed basis.

MANAGEMENT OF THE ONGOING PROCESS

Communications and Feedback

The communication and feedback process had improved greatly from the initial years to the time of the case research in 1986. As the Chinese and C-E worked together, the parties to the agreement developed greater trust and respect. In the early days, there was a lot of secrecy on the Chinese side. Often C-E did not get answers to its questions. Czapracki made the following observation:

> As we have had some people back here for training two or three times, and as we have been over there multiple times, and engineers get to know engineers, they

become more comfortable with one another and the job in general. They seem to be able to put aside the official secrecy, if you will. You will occasionally get a hint or innuendo of what the real problem is—a bureaucratic snafu, somebody forgot to order the material and it did not come in. We are getting more and more insight into what the real problems are. I think both sides are feeling more comfortable with that.

On several occasions, the Chinese engineers admitted that they did not know what to do in a certain situation and requested that a member of the C-E team make the decision for them. This was quite unusual in comparison with some of C-E's other customers in other parts of the world and particularly to customers in the United States. Bob Haun recalled,

I would be sitting in my office and would look up, and all of a sudden there would be four Chinese standing there. They would say, "Mr. Haun, you have to make the decision, we don't know which way to go." I tried at first to tell them that this was their decision, but this was one of the first things I found, they just hated to make decisions. But they loved you to make them and they accepted your advice.

They never came back to me once for making the wrong choice on their behalf, as many people told me might happen. Occasionally they came to me with two choices and asked me to recommend what they should do. If it was something I was very familiar with, I would do it right there. But I would usually call up our people, our engineers, and ask their opinion. Sometimes we would have a little meeting on it, or sometimes I would send a telex to China telling them we can do this or do that, and I have chosen to do this. If you have a problem with that, get back to me right away.

We had hundreds and hundreds of telexes back and forth, particularly to Shi Xiren in Shanghai. He was the head guy at the boiler works as far as we were concerned. He spoke seven languages and was a very brilliant guy. I think he cleaned latrines during the Cultural Revolution. The guy at Harbin Boiler Works was Wu Yiquan, the chief engineer. There was a lot of communication on this job.

Initially there was some concern that the engineers at C-E would not have sufficient feedback and access to information to monitor the progress of the Chinese plants. In order to better facilitate communication over the life of the project and also as a means of communicating about future projects, C-E set up an office in Beijing. The circumstances of this corporate office were described as follows:

The living room in the Friendship Hotel served as Joe Huang's office. He was a Chinese American and served as country director for a three-year commitment. He managed to help when things got tight on the communications side and when we could not get an answer out of the Ministry of Machine Building Industry. We could telex Joe or give him a call, and he could go over to the Ministry for us to make sure a particular letter got there. That was an advantage in the early days, having him there to confirm things.

He would also monitor the market, maintain harmonious relations not only with our licensee, but also with the Ministry of Machine Building Industry, the

Ministry of Water Resources and Electrical Power, and with the end users. His responsibility was not strictly related to our contract signing. His job was essentially to develop business with China, corporatewide. Recently, he has become more involved in other divisions of C-E, and the time he has to devote to us has diminished.

The Power Systems Group sent five people over to work in the Beijing office in October 1985. Jim Van Fleet, a dedicated oriental scholar, and his wife head this group with three other engineers, who are performing many of the functions that Joe Huang originally performed.

Scheduling and Performance

There were several key milestone activities that were to take place from the initial contract signing to the final erection of the boilers. C-E performance with each of these activities was evaluated based on how close to schedule C-E remained and how many complaints and concerns were voiced by the Chinese side.

C-E did the initial conceptual design in collaboration with the Chinese engineers in the United States. The Chinese side then did the detailing of the plant design in China. Before the contract had even been signed, C-E executives had submitted a proposal describing specific phases with time frames. As time passed, each phase became more detailed. The various disciplines, for example, the structural steel group, the pressure parts group, and fuel piping group, worked separately with Chinese groups to complete the tasks.

By January 1985, C-E had shipped everything except the coordinated controls, which C-E had subcontracted to Foxboro. The delay here was caused by design and contract changes initiated by the Chinese. These delays fell under the contingency clause defined in the contract and did not reflect poorly on C-E. At this time, the verification units were about two years behind schedule.

Overall, C-E executives claimed that the Chinese side performed their functions very well. Bob Haun made the following comment:

> I feel very comfortable. In many ways they are doing better than we are. One of the problems you run into in erecting a boiler is that you have to use the boiler structure to erect itself. You hang parts off beams, and as a result, you put stress on these members that are normal loads during operation. You must also have access to get pieces of equipment, so you must construct what they call "leave off steel," where you leave members out. We wanted to be sure that was all done properly, so we sent two fellows over there, a structural steel engineer and an erection expert superintendent. They came back and said that the Chinese have some techniques to erect the boiler that C-E really ought to look at.

Customer Service and Training

The customer service and training portion of the technology transfer under the relevant agreement was an ongoing process and was the most extensive and time consuming of all activities related to the project. The Chinese trainees who were selected to work on this project were found to be top quality. As Alex Sivas commented,

> They are excellent. We have not had any problem with shop or technician people, either. They are all very well behaved, and they know their business. They come here for refinements. They are very capable in integrating this new technology with the knowledge they already have. The problem is not in the capability of the people, but in the system that frustrates them. They learn, but whether they can utilize the technology efficiently is another story. Unavailability of material, transportation, and so forth, is very frustrating. You can teach them how to use a PERT chart, but what good is the chart if you cannot get the right material and the transportation on time to fabricate? You can make all the PERT charts in the world, and it won't do you any good. They need a lot of coordination.

The trainees were chosen on the basis of their engineering qualifications. Many of them had studied English in their high school days, but for the most part, English was not a strong second language.

C-E sent many of its own people to China to assist during the design phase at C-E's own expense. The Chinese institutes did not ask for it, but C-E executives felt they needed the extra help. The boiler verification units are equivalent to ten-story buildings made of tubing. Due to their complexity, C-E felt it was necessary to provide erection and start-up supervision to assist and consult with the Chinese on how to erect the units and in what sequence, to ensure the success of the project. Bob Haun estimated that C-E probably gave the Chinese twice as much help as they paid for. Whenever C-E officials made a trip to China to render "free assistance," they tried to schedule other business for efficiency's sake.

In late 1985 and early 1986, the Chinese factories were in the midst of constructing the two verification units. They had ceased sending engineers to the United States for training. It was speculated by C-E executives that they had either run out of areas to train or they had run out of qualified engineers to train. Many C-E executives involved with the project felt that the originally planned 800 man-months of training had been excessive. The Chinese were told this at the beginning, but insisted on that amount. C-E offered to trade man-months of training for man-months of technical assistance from C-E, but no decision had yet been made.

When the Chinese side told C-E they would not be sending people to the United States for a year and a half while they were working on production, they also indicated they expected to be getting a new crop of young engineers. The Shanghai authorities expressed their desire to give the new engineering graduates some experience first in their own shops before they started cycling them through C-E programs.

At the time of the case research in mid-1986, the last group of Chinese trainees had been with C-E from June to December 1985. From 1986 through mid-1987, the Chinese side needed all of their available engineers to produce and construct the verification units. Czapracki summed up his feelings with respect to the effectiveness of the training and allocation of man-hours used by the Chinese affiliates:

> The most effective training we feel has been done in the United States; however, there is provision in the license agreement for what we call technical assistance, some 70 man-months of our experts in China to help them with engineering, manufacturing, anything under the license agreement. There are no specifics as to what the 70 man-months will be used for—it is a matter of negotiation. We have used up about six man-months, a nominal amount up to this time. For the most part, the Chinese seem to be holding this purse of time very tightly. I wanted to send people over in the early days to help them with problems in the computer area. I had the right guy picked out, he was free, he had good working relationships with the trainees here, and they obviously had a need for this man but they would not send me an invitation for him. They were just not ready for him.
>
> Basically, they can tap into that time allocation anytime during the 15-year agreement. They have usually come to us and told us they would like either this expertise or this particular man based on some trainees' experience here. We have one engineer who is key as far as circulation studies go; part of the engineering design is directed to the water and steam side of the boiler, and they requested him specifically by name to go over there and help with their studies. When he came back he had lost 30 pounds. I sent him over without realizing he did not like Chinese food, so he had a tough time of it. He went to Shanghai, Harbin, and Dong Fang.
>
> I think they have come to realize, especially when it comes to these verification units, that it isn't so much the training as it is the guidance and supervision they need. They were told even beforehand: Why do you need 800 here and only 70 there? You have to put units together, this isn't going to be enough, and they made whatever decisions they did going into that negotiation and have never revealed their reasons why they came up with that number. I suspect it was arbitrary in their case, so they are finding now, as we have been telling them, that they are going to be short on these 70 man-months. They don't feel that once their 70 man-months are expended that they are going to be able to come up with the cash to easily pay per diem for a technician in China. It looks like they are saving it up for the most critical aspect of the boiler erection. We are a bit nervous and a bit concerned. We are putting our own people over there outside of the agreement to make sure our commitments are lived up to.

Financial Situation

The financial arrangements for the technology transfer were spelled out clearly in the initial contract. The contract stipulated a 15-year agreement with an up-front fee and no royalties. All the training was included, 800 man-months and 70 man-months of technical assistance. The Chinese side covered all miscellaneous expenses; C-E provided the technology.

In June 1984, C-E officials traveled to China to resolve price issues related to the changes that had been made in procedures under the contract by the

Chinese side. C-E was able to negotiate an additional $520,000 for interest, delays, and extra work that had been performed. At one point, delays on taking delivery of a control system by the Harbin Boiler Works resulted in a receivable of several million dollars for C-E. The Chinese managers had never heard of escalation clauses, nor did they understand the value of interest that could have been earned on that money by C-E.

If the Chinese asked C-E to do something special, they had to pay a special engineering fee. In 1986, C-E was doing several projects, one of which was to design a 600-megawatt lignite unit. This was not off-the-shelf information and therefore was paid for separately.

Assessment of Past and Current Position

Looking back over the technology transfer process that had transpired to date, Sam Blackburn summarized his philosophy as follows:

We only want a long-term relationship and always think in terms of a family relationship. There were times when, because of the overriding effect on long-term relationships, we might not have taken the best short-term approach.

When asked if the Chinese were short-term-oriented, Sam Blackburn commented,

I cannot read the Chinese mind that well. One of their common phraseologies is, "We are old friends." Old friends should imply, let's treat this relationship as though we have known each other for 50 years, and we want another 50 years. How much of that is real I don't know. There are certain of these people, for instance, the guy we deal with in Shanghai, who appear to be of a mind that long-term relationships are worth enough that they might concede some of their short-term interests. I'm not sure if some of the other people we deal with have that same attitude.

In thinking back over his experience in China, Sam Blackburn was asked to synthesize some of the main lessons he has learned and would want to tell others entering China for the first time. He responded,

Patience, patience, patience. I think by nature, none of us have been taught to sit back, relax, and wait. We have always been an offensive group of people. Move quickly, solve quickly, and get out. That cannot be done in China. If you go there locked into that attitude, you will never, never be able to do business with the Chinese.

Licensing has always operated under a family tradition which is a much more people-oriented relationship than a standard businessperson would ever think of developing. Most of the negotiating team had a licensing background of some kind, so, as far as the Chinese were concerned, we were more people-oriented than our competitors were.

Czapracki commented on the mistakes he felt he had made:

Mistakes? I guess probably not listening to the Chinese well enough, not being friendly enough, not smiling enough. Because of the deadlines, I am quite a factual person, and I never let the emotional side get in the way, and I can look back in retrospect; I pushed some of the people quite hard, and I think I could have been much more polite and still done the same job.

C-E's top management was known to be supportive of the China project, and there were never any major hurdles or pressures from the side of C-E management throughout the whole process. C-E executives indicated there was never resistance from people saying C-E was giving the technology away too cheaply, no innuendo about communists, and no ethnic harassments. There were friendly complaints that some people were double scheduled due to the training schedules, but overall there were no bad feelings or resentment on the part of C-E management or staff.

The first 300-meagawatt unit was scheduled to come on line in March 1987 in the Shanghai area. The 600-megawatt unit was being built by the Harbin Boiler Works, and in spring 1986 was still under construction in Ping Wei.

The Chinese side opened up enormously as the years passed. C-E was able to get statistical data concerning where the current power plants were and what their capabilities were. This information was previously unavailable. It was still difficult, but possible, for C-E officials to visit other power plants in China. These plants were considered strategic for the country, and Chinese officials were normally cautious about giving out information. Visits to power plants throughout China had provided a great deal of information for internal studies related to future joint ventures.

C-E did not feel it would be competing in China with its other international licensees due to the huge backlog of orders that licensees had. China had the capacity to utilize all the new power that came on line. They were using their local power plant design and construction capabilities to the limit and were importing whole units in addition to fill the need in the country.

FUTURE PLANS

C-E executives expected that they would be able to undertake more joint projects in China in the future as a result of the track record and reputation they already had and continued to develop. C-E expected the next step would be a joint venture with the Boiler Works affiliates in Shanghai or Harbin, either of which would be a good partner. C-E considered itself fortunate to have been the first to establish relationships with these two works. Harbin was the biggest, and Shanghai the second biggest boiler works in China. Beijing Boiler Works, the third, had become affiliated with Babcock & Wilcox, a competitor of C-E.

Long-Range Ambitions for China

C-E, a diversified engineering company, was involved in discussing numerous ventures with China, including equipment manufacturing for oil and gas exploration. C-E had expanded its presence in China by opening offices in Shanghai and Hong Kong, in addition to the existing Beijing office. Some of C-E's affiliated companies in the areas of pulverizing, simulated design, and computer-aided process control had already concluded technology transfer agreements. In total, C-E had ten technology transfer agreements at different stages in China by mid-1986. Of these, eight agreements were signed, and two were still in the process of negotiation; of these cooperative projects, two were joint ventures.

According to Joseph Condon, corporate vice president, these various activities were ongoing, and "at different times, they get hot and cold." Many of C-E's future strategies and plans were expected to be based on the foundation C-E had already established in China. China was high on C-E's list of priorities, even though it was viewed as a slowly developing and demanding market requiring unique methods.

Condon made the following comment with respect to C-E's future role in China:

> The Chinese are realizing that having a design is not enough. What they are really lacking is the manufacturing expertise and management techniques. So now they want joint ventures, a full commitment that we will support them and help them out.
>
> We see ourselves diversifying and increasing our commitments, making long-term commitments. We now have three of our groups telling us they need additional resources there to pursue their business on a daily basis with the Chinese, the MMBI. The question that I am struggling with is how do we do that in an efficient and cost-effective way for the Corporation so it gets a return on its investment. We know that the Chinese will absorb everything that you make available and will keep asking for more. We have Chinese groups here learning management skills. How much can we do, that is, how much can we get out of that market? That is the question I am struggling with right now. I have some ideas on how to do it, but I haven't yet resolved them. We have to really be careful.
>
> Is this a $300 or $500 million market for C-E for the next three years? Once we determine the size of the market, we have to determine what resources we can commit and how we can structure them. In these joint ventures, they want us to provide 20 engineers; they have to be people with practical experience. These are valuable people to us, not fresh grads out of MIT.
>
> I was down in Houston the other day talking to some of our people while a Chinese delegation was there visiting. From that visit 20 fundamental questions arose regarding a new joint venture. How do we price the raw material? How do we price Chinese labor versus U.S. labor? How do we manage? How do we determine profitability to the company? These are really fundamental questions with which the Chinese are only beginning to struggle.

EXHIBIT 1   Combustion Engineering: Key Events

Year	Event
1970	Used Business International as sounding board for selling into China
1974–1975	Used consulting services for market studies
1977	Established mail contact with China Technical Import Corporation*
1978	C-E technical seminar in Beijing; visited Harbin and Shanghai manufacturing plants; United States normalized relations with China*
1979	Submit proposals to MMBI; C-E visited by Chinese Nuclear Energy Society; Harbin and Shanghai plants surveyed by C-E; MMBI visits Windsor (Connecticut), Chattanooga (Tennessee), Monongahela (Pennsylvania), Chicago (Illinois), St. Louis (Missouri), Wellsville (New York), and Marion (Ohio)*
August	Negotiations started for technical assistance and prototype units (300- and 600-megawatt)*
1980	
November	Contracts and technical assistance agreements signed on November 21*
1981	
January	Mandatory deadline for technology transfer, standards, manuals, and general specifications
March	Contract effectively begins
March–May	Shipment of all documents to MMBI
September	Training begins; nine Chinese engineers arrive; some renegotiations on original contract
1982	
June	Most of engineering and R&D training related to the verification units complete
November	Design engineers begin 18 months of training
1984	
June	C-E officials negotiate additional $520,000 for changes in contract for interest, delays, and extra work performed
1985	
January	Everything shipped to China except coordinated controls from Foxboro
June	Last group of trainees come to United States for formal training
October	Power Systems Group establishes Beijing office with five people
1986	Last group of trainees returns to China
1987	
March	300-megawatt verification unit scheduled to come on line

*Items with asterisks are pivotal events.

CASE 8

Combustion Engineering: The Chinese Perspective

INTRODUCTION

This case was based on an interview trip to China in June 1986. The research team visited the Ministry of Machine Building Industry (MMBI) in Beijing, the Harbin Boiler Works, and the Harbin Power System Engineering and Research Institute in Harbin, Heilongjiang Province.[1] In Shanghai, we met with representatives of the Shanghai Boiler Works, the Shanghai Turbine Works, and the Shanghai Power Plant Equipment Research Institute. The purpose of the meetings and interviews in China was to investigate the Chinese side of the 15-year licensing agreement signed in November 1980, to become effective March 5, 1981, between the MMBI and the Power Systems Group of Combustion Engineering. This agreement provided for the transfer of boiler technology for 300- and 600-megawatt fossil-fuel power-generating stations to two provinces and Shanghai city. Shanghai was the recipient of the 300-megawatt technology; Harbin in Heilongjiang Province was the recipient of the 600-megawatt technology. Sichuan Province was to participate without a specific recipient task, having been assigned to the Dong Fang Company, which sent some of its personnel for training under the agreement.

---

[1] In December 1986, the MMBI was merged with the Ministry of Ordnance Industry to form the Machine Building Industry Commission. We shall continue to use the name MMBI in this case study.

Discussions were held with officials of the MMBI who had participated in negotiations leading to the licensing agreement and in its implementation and also with the leading technical and administrative personnel at the factories and research institutes in Harbin and Shanghai mentioned above.

DEVELOPMENT

Wu Wufeng, commissioner and secretary general of the State Science and Technology Commission (SSTC) foresees immediate emphasis being placed on energy, transportation, communications, and raw material resource development. The current five-year plan calls for 4,000 to 5,000 megawatts of electric power production to be added annually, but the demand is still higher. The Chinese would like to use the latest technology, but they realize this desire must be balanced against economic constraints. Additionally, the Chinese would prefer to use already proven technology versus technology still in the experimental stage to ensure consistent progress. As various technology transfer opportunities arise, they are evaluated based on price, sophistication of technology, overall benefits derived, and social impact. Turn-key projects continue to receive less attention than technology acquisition and absorption, which receive the main focus. These factors led the Chinese to C-E's boiler technology as part of their socioeconomic development plan.

Why C-E Was Chosen

According to Wu Yiquan, chief engineer at the Harbin Boiler Works, after the decision had been made to import boiler technology, Chinese engineers and ministry officials reviewed the world competitors involved in this technology. Chinese delegations were sent to Europe, United States, and Japan to determine who should be given the contract. C-E and Babcock & Wilcox were the U.S. companies under consideration.

C-E was chosen primarily because it was the original developer of boiler technology and, over the years, had achieved a reputation for technical excellence. A friendship and common understanding had also developed between C-E and Chinese officials, which made a big impact on the Chinese government's final decision to use C-E technology. Other factors considered important in making the decision included *quality* of the technology, *price,* the *economics* of negotiating a suitable arrangement, *feasibility* of transfer, and *suitability* of the technology. All of the boiler companies being considered and their products were well known within the industry and to the Chinese.

NEGOTIATIONS

The Chinese felt many of the problems encountered during the negotiations came from the U.S. side. The main obstacle they cited was the abundance of terms and conditions imposed by U.S. law. The lawyers would compare the conditions of the contract to a "yellow book" and would continually say "No, no, no, and, we need more money." The Chinese felt the Japanese and EEC companies were much more flexible and able to negotiate because they did not bring lawyers.

The Chinese offered two reasons for the lengthy two-year period to negotiate the contract with C-E. First, it takes time to obtain approvals from various PRC government officials. Second, it takes time to become acquainted with the U.S. company before any agreements are signed, a necessary process to the Chinese.

START-UP

Paper Transfer Issues

After the contract had been signed, Chinese engineers wanted full documentation on all manufactured products. However, the documents they needed the most were delayed and the less important documents were the first to arrive. The Chinese felt that C-E management had allocated very little manpower to delivering the documentation and drawings. They also generalized that C-E's implementation did not always follow the terms of the agreement.

The most troublesome aspect of the document transfer experienced by the Chinese dealt with the translation and conversion of dimensions after the drawings arrived in China. Technical terms, equations, and descriptions had to be translated, and dimensions had to be converted to metric standards. Since U.S. methods of projection in engineering drawings are different from those in China, numerous adjustments had to be made. The comment was made that U.S. companies never change the documentation to conform to Chinese specifications, whereas most EEC countries do this work before transferring the paper. During the negotiations, the translation of documents was not clearly understood nor was it discussed. It has since been considered one of the disadvantages in selecting a U.S. company for technology transfer.

Training Programs

The 300- and 600-megawatt units built in Shanghai and Harbin, respectively, consisted of two separate contracts: one for the boiler technology to be licensed from C-E, and one for the turbine and generator technology to be licensed from Westinghouse. The Chinese compared the method of training provided by each company and felt C-E's on-the-job training and participative design were more suitable than Westinghouse's classroom-type training. Chinese engineers dispatched to the United States were experienced designers, and this made simultaneous design and on-the-job training successful.

The training was divided into three phases as follows:

1. Proposal design phase lasted three months. C-E engineers completed this work while Chinese engineers observed.
2. Engineering design phase lasted one year and was completed by Chinese engineers under C-E supervision. Sixteen Chinese engineers were sent to the United States for this stage.
3. Detailing phase lasted one year and was carried out totally by the Chinese engineers in China. Twice during this phase, three to four C-E engineers went to China for two to four weeks to oversee the Chinese engineers' work.

MANAGEMENT OF THE ONGOING PROCESS

According to the terms of the agreement, C-E would provide people for start-up, adjustments of the first boiler unit, and after-sales service. In addition, the Chinese have been trying to upgrade their management standards and are engaging the help of C-E to help them reach this goal. One engineer from Harbin was assigned to come to the United States for management training, and several C-E people were assigned to Harbin for quality assurance and project management. Both C-E and the Chinese agree that expertise in management and service is more difficult to transfer than technical documentation and expertise related to design and component hardware. However, management expertise is essential if the on-going process is to be effective.

Components Sourcing and Manufacturing

There are many differences between the standards and sizes of machine tools used by C-E and those of the corresponding factories in China. These differences had the potential of causing tremendous problems in both design and construction of the power plant. Since Chinese factories can only forge certain size parts, they have limited their forging activity and have concentrated on casting as their preferred method. Forging, on the other hand, is the preferred and most common method used in the United States. Without careful forethought and planning, these different methods of manufacturing make component interface very difficult.

The difference in standards used around the world caused the Chinese to adopt the U.S. standard for many of the components used in the initial verification units. The similarity between U.S. and Japanese standards was another reason for choosing the U.S. standard. To address the problems of sourcing in China, C-E engaged in a joint design effort, wherein Chinese standards for widely used components could be introduced during the planning process, thus saving time and effort in later stages. When small quantities of parts were required, China would usually buy direct from the United States rather than try to source them locally.

Quality Control

Contracts between C-E and the Chinese provided for performance tests specifically agreed upon by both parties. Since the 300-megawatt verification unit was completed in August 1987 and successfully tested, both C-E and the Chinese felt that this aspect of the technology transfer was successful. In addition to the performance testing of the physical plant, the Chinese also evaluate and judge the soft technology (software) and the management expertise, but did not express an opinion on this issue.

Chinese officials found that C-E's technology was very advanced and, as a result, it was costly. Since the cost issue has caused some difficulty with the final customer (MWREP), efforts have been made to lower costs whenever possible. Most of the major parts and components have been locally sourced and are manufactured at the Harbin Boiler Works with only a few subcontracted to other Chinese suppliers. This dramatically lowers the costs and allows Chinese factories to produce and be responsible for the quality they desire.

Communication and Feedback

The Chinese felt their communications with C-E were very good. Telex machines have been used extensively throughout the course of the project with certain communication protocols developing over time. The Chinese attributed this good atmosphere between parties to frank negotiations and open communication from the very beginning. Occasionally, Chinese engineers found quality problems in C-E–supplied components, which had to be repaired at the Harbin plant. There were also some defective components sourced by C-E from other U.S. suppliers. As these problems were discovered, prompt notification was made and corrective measures were taken.

The transfer and retirement of C-E personnel involved in the China project required that contact personnel also be changed. C-E was considered very good about informing the Harbin Boiler Works of these changes to minimize the negative effect of having new people emerge on the scene after close working relationships had already been established. Both old and new personnel went to China to make introductions and transfer any knowledge necessary to ensure the uninterrupted flow of the project.

Evaluation of Experiences

Chu Pinchang, chief engineer at Harbin Power System Engineering and Research Institute, made the following comment:

When we look back on the contract, it seems it has been implemented smoothly. Both sides profited from the technology transfer project. China obtained technology and components from C-E, and C-E earned a profit in dollars and was paid for the licensing of its technology.

There are other long-range advantages that C-E and other U.S. companies have received. C-E is able to sell parts to China on a continuing basis. More important, C-E has been introduced to the Chinese market and has made many important contacts with Chinese officials that will be important for continued business relationships. On the technology side, imports will probably be increased in the future as a result of the good work completed to date.

After the contracts had been signed, the Chinese felt that C-E's price for its technology was reasonable in part because of the intense competition between the various U.S. and foreign companies that bid for the project. China benefited greatly by cutting the time required for catching up with advanced technology and world standards. The nature of the dynamic technology transfer also provided to China a continuous updating of the technology for at least a 15-year period and most likely beyond.

FUTURE PLANS

The Chinese side believed that the relationship with C-E will last for a long time. Work has already begun between C-E and China on a third power plant to be built in China and financed by the World Bank. Both the Shanghai and Harbin Boiler Works have sent a design team to C-E to undertake a new boiler designed specifically for meager coal and lignite. Over the long term, China plans to export the Harbin 600-megawatt design unit, initially to Southeast Asia and to other less-developed countries. Since labor costs are low in China, C-E is very willing to cooperate in an effort to produce and sell a more competitively priced product. China expects to co-produce and jointly market with C-E a product that is particularly competitive with the Japanese, who currently produce a much less expensive product than the United States.

The Chinese feel very good about the cooperation they have received from C-E and intend to cooperate only with C-E in the future for this product. Even though the 15-year contract term is very long, the Chinese expect to renew the agreement when it expires. The Chinese chose to establish and maintain their relationship with C-E (versus the number two U.S. competitor, Babcock & Wilcox) because C-E is more experienced, has more worldwide licensees, has a better reputation, and offers a slightly better price than its competitors.

Wu Yiquan, chief engineer at Harbin Boiler Works, indicated several things that C-E can do to ensure successful cooperation with the Chinese in the future: (1) lower its prices to remain even more competitive, and (2) increase the scope of its management training.

CHAPTER 3 _____

Comparative Assessment

Our comparative assessment of the four case studies will be organized by the stages of the technology transfer process, in accordance with the outline we adopted for our case studies (see Appendix). For each phase, we discuss the strategies of the four companies investigated and we assess their performances against criteria for success that have been previously developed (Von Glinow, Schnepp, and Bhambri, in press). In addition, we shall consider some overall strategies and success criteria that span the entire process or cut across several phases.

OVERALL STRATEGIES AND SUCCESS CRITERIA

As we have seen in Chapters 1 and 2, there are a number of criteria that U.S. firms wish to optimize as they enter into a technology transfer to China. Of the most salient of these overall indicators, we discuss strategies concerning (1) profitability and foreign exchange management, (2) information exchange and continuing feedback, and (3) penetration of the Chinese domestic market. We further consider suitable criteria for success under each of these rubrics.

The four companies in our study pursued strategies that differed in a number of respects. Foxboro entered the China market with an aggressive strategy and negotiated a joint venture very early in the Open-Door period. Although the 1979 Joint Venture Law had already been enacted, implementing regulations had not yet been published at the time.[1] As a result, the principal Foxboro negotiator had to ensure that all conditions were specified in detail in the contract. In this sense, the project resembled a cooperative venture, according to the classification described in Chapter 2.

The other three companies, Cummins, Westinghouse, and Combustion Engineering, used a more defensive strategy of licensing in their initial entry into China. The risks associated with this style of affiliation are clearly more limited but also the profitability is usually thought to be lower than that obtained in equity joint ventures or wholly-owned subsidiaries.

Profitability and Foreign Exchange Management

Since Chinese currency (renminbi) is not freely convertible, foreign exchange management has been a critical issue for all foreign-invested enterprises in China. The part of revenues received in hard currency or the part of earnings of a joint venture that can be repatriated to the United States is high on the list of factors determining the profitability of a commercial venture. Thus foreign exchange management becomes a critical factor for each firm.

Foxboro received technology transfer fees (royalties on sales) in the form of hard currency from the joint venture company, but Foxboro paid for its share of rent for land in U.S. dollars. The joint venture reported profits beginning the second year of its operation, but it lacked the foreign currency to buy necessary foreign-made components for its products. As a result, Foxboro had to make further investments to support the joint venture activity. The local currency profits were reinvested, resulting in tax advantages. In the short term, Foxboro profited from revenues generated by the sale of technology and components intrinsic to the technology transfer project and by direct sales, which were promoted by the visibility gained through Shanghai-Foxboro Company Limited (SFCL).

SFCL's foreign exchange management problems were eventually eased by a number of methods. The U.S. company, after several years of SFCL operation, bought back parts manufactured in China. Since about 1985 the joint venture was authorized to sell its product for partial payment in foreign currency to agencies in China having hard currency income. SFCL also, reportedly, obtained dollar allocations from the Shanghai government and recently has been able to buy foreign currency at the Shanghai Foreign Exchange Adjustment Center (Simon, 1987) to help satisfy the joint venture's requirements.

The dollar obligations of SFCL included payment of part of the salaries for the two American expatriates in residence in Shanghai. It is not known if the parent company has repatriated any of SFCL's profits as of this writing. However, Foxboro experienced gains in direct sales in China outside the joint

---

[1]The implementing regulations were published in 1984.

venture, which was attributed to the presence of the company and publicity surrounding the joint venture.

Cummins also received payment for technology transfer in the form of royalties tied to the production volume of the Chongqing licensee. In addition, Cummins was paid in foreign currency for the kits and parts supplied to the Chinese affiliate. At a later stage, Cummins began to buy back engine components manufactured by Chongqing Automotive Engine Plant, which again allowed the Chinese factory to expand its production volume and thereby to increase Cummins Engine Company's earnings. The project was further aided by the technology recipient's preferential status for foreign currency allocations as recipient of awards for successful technology absorption. Finally, Cummins was eminently successful in expanding direct sales in China because of the company's visibility in the country. In fact, for some years into the licensing agreement, Chinese purchasers preferred the U.S.-made engines to those manufactured in Chongqing.

Both Foxboro and Cummins established Chinese-owned service companies that promoted direct sales of the U.S. parent companies and received commissions in U.S. currency. It was originally hoped that the infusion of hard currency by the U.S. firms would ultimately benefit their business activity. However, at least in the Cummins case, the service company was required to surrender its entire dollar receipts to the government. Cummins, like Foxboro, eventually bought back parts manufactured in China, thus improving the hard currency balance of the technology recipients and enabling them to purchase more U.S.-made components. In January 1987 it was reported in the Chinese media that SFCL's foreign exchange transactions were in balance for the year 1986, indicating that the most burdensome problem of the joint venture had finally been solved. However, Simon (1987) reported ongoing difficulties that lacked long-range solutions. Cummins has found its commercial association with China profitable and has expanded its licensing activity to a second and significantly larger producer of diesel engines in Hubei Province.

Westinghouse and Combustion Engineering received payment in cash for their technology through a U.S. EXIMBK loan and received revenues from the sale of components; the licensees' product contained about 70 percent and 30 percent foreign-made parts, respectively. Westinghouse executives complained, however, that the project, on balance, was not sufficiently profitable to satisfy the corporate management. The U.S. company had to increase the consulting services at its own cost to support the project and to uphold Westinghouse's reputation. On balance, it appears that Westinghouse expected a higher level of long-term profitability (six years) and curtailed its China operations in the power-generation sector when these expectations were not fulfilled.

Technology transfer turned out to take longer than anticipated, and the local industry could not satisfy the increasing demand for electric power-generation equipment. In response to this pressure, the Chinese government revised its strategy and bought turnkey power-generation plants from abroad, but Westinghouse did not find it profitable to change directions and bid for such Chinese contracts.

Combustion Engineering, on the other hand, succeeded in expanding the company's activities in China in the course of the years since the inception of the project, indicating the company's willingness to adopt a long-range view and to invest in the expectation of larger profits in the future. The company has developed an international network of licensees and is oriented toward accommodating foreign affiliates. Handling all China projects for the good of the company and increasing their profitability have, however, remained major concerns of Combustion Engineering.

Information Exchange and Continuing Feedback

It has long been recognized that language and cultural barriers to information flow can decrease the efficiency of technology transfer (Kedia and Bhagat, 1988). It was concluded by all four U.S. companies that feedback was substantially enhanced by a local presence. Foxboro, as a partner in a joint venture with resident expatriates, did not suffer from this problem. However, Cummins, Westinghouse, and Combustion Engineering all found that the feedback from the Chinese side was inadequate for monitoring the transfer process, particularly in the initial stages of the project. Even during on-site consulting visits, Cummins engineers had limited access to information concerning the progress of the technology transfer process. However, this problem progressively diminished over time, presumably as Chinese self-confidence grew and as personal relations developed, until the flow of information reached satisfactory levels. The personnel of the U.S. company reported that the Chinese affiliate's personnel were at first quite aggressive about voicing their criticism of shortcomings on the side of the technology supplier, but were reluctant to admit their own difficulties.

Cummins managers believed that they could have been of greater assistance in furthering the transfer process if they had been taken into their affiliate's confidence more fully. As late as 1985 the American technical managers believed that they had insufficient information to make reliable judgments on the level to which the transfer process had progressed, but considerable progress was recorded during the subsequent two years. Cummins also believed that communications would have been improved if the U.S. company had had a presence in China, even if not at the actual site of the Chinese technology recipient. Cummins only opened an office in Beijing in 1984. Westinghouse managers also said that their company's offices in China contributed substantially to the communications between them and the technology recipient.

Combustion Engineering managers repeatedly had the experience of being asked by Chinese engineers to help them make decisions or even to make decisions for them. In the Americans' experience, their Chinese counterparts never complained about having been given poor advice after the fact, as was initially feared by some of the technology supplier personnel. This experience underlines a basic managerial problem encountered in China and the need to remedy it.

Combustion Engineering also established an office in Beijing to facilitate communications. At first, this office was manned by a Chinese American, but later the Power Systems Group sent a Caucasian American who had deep interest in Chinese culture to head the staff. The new office manager proved to be of great help to the company. One of the U.S.-based managers said that he had made great efforts to promote the technology transfer process, but still felt that he could have done more and, in particular, should have shown more patience and been more polite. He concluded that progress would not have suffered if he had pushed people less hard. These reflections emphasize the great efforts required for successful transcultural technology transfer.

Chinese Market Penetration

The penetration of the Chinese domestic market is an important aim of most foreign companies. Access to domestic markets is controlled by the Chinese government, and any gains in this direction are important success markers. There are two levels of market access, depending on the location of the source of the product inside China or abroad. In most cases, technology transfer projects enjoy the support of the government since their approval in the first place is usually based on the domestic demand for the product. As a result, market access for locally produced commodities is likely to be achieved. Direct access for the sale of foreign-manufactured products is harder to achieve since it necessitates foreign currency allocations by the appropriate authorities and therefore represents a higher level of achievement and success.

Shanghai-Foxboro's products have been successfully marketed in China, and demand for the industrial control instruments has exceeded the joint venture's production capacity. However, sales have been limited by the joint venture's ability to cover the foreign currency costs of U.S.-manufactured components, as discussed above. Also, Foxboro succeeded in widening direct sales. On some occasions Foxboro won contracts in China even though its bid was not the lowest. This success has been quoted as evidence for the promotion of the parent company's products due to the visibility of the company through SFCL.

The engines produced by Cummins' Chinese affiliate eventually became successful in China. The affiliate had to overcome initial customer resistance originating from the poor reputation of an engine produced by the Chongqing factory from technology acquired from France in the 1960s. Again, direct sales in China which accompanied the technology transfer activity of Cummins have been profitable for the U.S. company. The Chongqing affiliate's managers said that they did not see any problems arising from the competition of their U.S.

affiliate in the China market. Like Foxboro, Cummins established a Chinese-owned service center, and both companies promoted their direct sales through the service organizations by offering commissions in foreign currency. This incentive-driven marketing strategy turned out to be successful and was bolstered by the availability of servicing and spare parts through these local companies.

The product based on the technology transferred by Westinghouse and Combustion Engineering was of a considerably larger scale. It took five to six years to produce the "verification units," and therefore their cases cannot be considered as comparable with those of Foxboro or Cummins in terms of market penetration. It is, however, very clear that China has continued to suffer from an acute shortage of electric power–generation equipment, and, in fact, the country's modernization and development programs have been power limited. The problem here has been the failure to transfer technology and to expand production fast enough to enable the Chinese plants to satisfy the demand. In fact, China has, in recent years, purchased turnkey plants from foreign suppliers in preference to projects proposed by Westinghouse, which were based on the technology already transferred and included partial manufacture in China. Westinghouse, together with Combustion Engineering and Ebasco, formed a consortium with the Chinese factories and bid for the sale of power plants in China (Ningpo, Jiangsu Province) and abroad (in Egypt, for example). Combustion Engineering has continued to expand its activities in China, but Westinghouse has recently decided that the market for power-generation equipment has not proved sufficiently profitable for the company to pursue activity in this product line in China.

STAGE ONE: DEVELOPMENT

During this preformative stage the foreign technology supplier is concerned with acquiring information on Chinese industry and markets; thus the relevant strategies of each firm are examined and their performance is evaluated using appropriate criteria for success.[2]

Early Information on Chinese Industry and Markets

The criteria for the success of foreign business ventures in China are not standard, and the evaluation of many conventional indicators cannot be accomplished through traditional methods. A major criterion for an American firm's successful operation in China, as elsewhere, is the level of access to information on local conditions. However, in China the barriers to information-gathering and to its verification are particularly formidable (Mayer, Han, and Lim,

---

[2]See Appendix on page 255 for an outline of the four stages of the foreign technology transfer process.

1986). As a result, the recipient's level of willingness to provide access to its own plants and to mediate access to the installations of its suppliers and of potential customers represents a major indicator of progress toward a successful relationship. Further, the provision of such access and openness to discussion of technical and administrative difficulties on the Chinese side indicates a level of trust between the affiliates, a basic and essential condition for a fruitful future. Additional areas of openness indicating progress include plans for future development and the role of the Chinese company in the development plans of the region and of the country.

Foxboro Company. A British subsidiary of Foxboro had begun about 1975 to participate in exhibits organized in China, and ethnic Chinese personnel from the company's Singapore office participated and facilitated communication. Information on China collected over the course of these visits was available to the parent company when the time came to consider a partnership with China. Subsequently, Foxboro received all its information concerning the industrial controller industry and market through its Chinese counterpart. Thus, Foxboro's information during this background stage was only as good as the Chinese estimates. Ultimately, Foxboro had little accurate market data of the true demand for its product, although the Chinese were obviously keen on Foxboro's technology. Ministry of Machine Building Industry (MMBI) officials arranged visits for Foxboro personnel to factories and to their customers, in addition to allowing full access to the plant where the joint venture was to be established. Market demand estimates and choice of technology to be transferred were entirely based on the information gathered in the course of these visits.

Cummins Engine Company. Cummins Engine Company's then chairman of the Board, J. J. Miller, visited China as a member of a large U.S. trade delegation in 1975 (prior to normalization of relations between China and the United States), and at that time acquired a diesel engine manufactured by the Shanghai Diesel Engine Works. Analysis of this engine gave Cummins clues concerning the level of the Chinese diesel motors industry. However, no additional research or planning was carried out at the time. In 1978, when serious discussions began, the U.S. company had minimal information on Chinese conditions and received only limited input through its Chinese counterparts, a bureau of MMBI and its negotiating agent, Techimport. As it turned out, the market information obtained through these sources was quite faulty; the early estimates put the demand at 26,000 units per year, and this figure was reduced substantially during the negotiation process to 6,000, and then even further. In 1985, Cummins personnel told the researchers that even then the market demand was still poorly understood.

During the early years of contacts, Cummins personnel felt frustrated by their inability to penetrate the Chinese screen of confidentiality, and they returned from inspection tours and negotiating meetings with the feeling that their information base was very limited. Manufacturing experts were, however, taken to visit selected plants and were invited to contribute to plans for the expansion of the designated technology recipient, the Chongqing Automotive Engine Plant. Cummins, as a result, had information on the level of Chinese technology (the industry was found to be outdated by 20 to 30 years), and the technology supplier understood the role planned for the technology recipient as a trial base for the absorption of foreign diesel engine technology.

Westinghouse. At Westinghouse, a Chinese American physicist collected background information from Hong Kong publications and thereby provided Westinghouse with a body of independent information on China's power-generation equipment industry. This resource supplemented the exposure provided company personnel during fact-finding visits organized by the Chinese side (also MMBI). Also, Westinghouse had had a very early presence in China. Prior to the founding of the People's Republic in 1949, Westinghouse had trained numerous Chinese engineers, and as such, fell into the category of "old friend" in 1978, despite the absence of formal contacts in the interim period. On the heels of President Nixon's visit to China in 1972, then Board chairman Donald Burnham encouraged Westinghouse executives to penetrate the China power-generation market. Ultimately, it can be stated that Westinghouse was successful during this stage of the project.

Combustion Engineering. Combustion Engineering received early information on China through Lummus Company, an engineering subsidiary, which planned and erected ethylene plants in 1973 and 1978. The parent company realized the market potential of China for industrial boiler technology and was aware of the successful entries of some of its European licensees into China. Beginning in 1978, Combustion Engineering began official contacts at the invitation of an MMBI bureau, but also found that most meetings were one-sided, with the Chinese side keeping information under strict control. Eventually, the U.S. company personnel were exposed to the conditions prevailing at the designated technology recipient plants in Shanghai and Harbin. Combustion Engineering was reasonably successful in collecting information, given the considerable barriers it encountered.

STAGE TWO: NEGOTIATIONS

General

Since the development of trust and confidence toward the foreign company and its personnel represents an important aim of the Chinese side during the initial phases of the project, the atmosphere or climate during the negotiation sessions provides indications of progress toward this aim. Relaxed discussion, an absence of tensions, and a lowering of the level of formality are all positive signs. On the other hand, some foreign companies contend that it is very difficult to conclude an agreement without some open rifts or even confrontations.

Foxboro Company.   The lead Foxboro negotiator, a corporate vice president with rich international experience, was successful in keeping the climate free of rancor during meetings. When progress seemed blocked, he simply informed his Chinese counterpart that he would return home and await further ideas from the Chinese side as to how to further the negotiations. The attitude adopted by the U.S. company was that it would be good to reach an agreement, but this was not essential and might well have been out of reach. This approach resulted in a lengthy but calm process that eventually culminated in the successful establishment of the joint venture. Cummins could afford a relaxed attitude, since sales to China continued throughout the negotiations and were independent of them.

Cummins Engine Company.   Cummins had a particularly difficult experience, since the negotiating agent on the Chinese side was China National Technical Import Corporation (Techimport), a trade corporation subordinate to the Ministry of Foreign Economic Relations and Trade (MOFERT). Also the team was made up entirely of nontechnical personnel who did not have a direct stake in the venture. The U.S. team members said that the interaction sometimes resembled shadow boxing, since the Chinese team did not sufficiently understand the significance nor the content of the proposed technology transfer venture. The Chinese sometimes adopted, what seemed to Cummins personnel, a confrontational stance, and negotiations did not progress smoothly. At one point there was a real crisis, but when the ethnic Chinese engineer was brought in to interpret and mediate, the tensions eased. The U.S. side also used confrontational tactics and broke off negotiations when a high-ranking corporate official in attendance decided that the Chinese position was unacceptable. During the last session, when the final financial conditions were to be agreed upon and the same executive was present, the agreement was signed following another confrontation. It seemed at the time that Cummins had won but actually, some months later, the company had to make concessions under pressure from the Chinese central government agency (an MMBI bureau) responsible for the venture.

Westinghouse.   Westinghouse faced the greatest challenge of all the companies studied since it came in to bid when other competitors were already further advanced in the negotiating process. In fact, by that time General Electric was assumed to be the winner. However, because of patience, perseverance, and flexibility, Westinghouse eventually won out over G.E. The climate was generally good during the lengthy negotiations process, but there were some differences concerning the limitations on the technology to be transferred and the U.S. company insisted on strictly adhering to the scope of the agreed-upon products.

Combustion Engineering. The Combustion Engineering negotiator avoided confrontations by discreetly informing his opposite team leader that he was forced to return to the United States when negotiations bogged down. By this stratagem, demonstrative walkouts were avoided, but the aim was achieved: The Chinese side undersood that it had to come up with new proposals before further progress was possible. Actually, Combustion Engineering negotiators lived through some difficult times. The Chinese side negotiated simultaneously with one or more competitors and tried to play off one supplier against the other. One U.S. representative said that they never knew where they stood. Also, at one time the agreement was almost ready, as far as the Americans understood the situation, when the Chinese side added a new group of commercial personnel to its team and insisted on beginning to negotiate the agreement once again from the beginning. The final contract was much more detailed and about three times the length of the original document. It is concluded that the Combustion Engineering team was subjected to a great deal of stress but managed to preserve decorum and used diplomacy rather than confrontation.

Chinese Decision-Making Process

The penetration of the protective screen surrounding the decision-making process and the identification of decision makers represent key success factors during the negotiations phase. Success in gaining an understanding of the Chinese negotiating team and of all personnel present in the room can provide important clues concerning the secondary stakeholders. A further step would be actual contact with the leading personalities and the opportunity for lobbying them, even though the effectiveness of lobbying a decision maker may often fall short of expectations based on U.S. models. The decision-making power, or the willingness to accept responsibility for decisions, is often limited, whereas strong reliance on consensus is common, even among officials of reasonably high stature. The appearance of ranking "leaders" at social functions is certainly significant and also provides opportunities for initiating some level of management of the Chinese decision-making process. In particular, the level of the bureaucratic hierarchy's representation during a visit by a senior American executive and the level of publicity given to such a visit provide important data on the success of the interaction.

The Chinese decision-making process is time-consuming because it is based on consensus-building, requiring the participation of a number of stake holders. Many foreigners are impatient and resent the amount of time required for negotiations. The management of this problem is important for foreign companies, since otherwise the investment of personnel time becomes excessive.

Foxboro Company.   Foxboro coped very successfully with the time factor during negotiations. The lead representative spent limited periods in China and returned home when negotiations lagged and progress was blocked. He returned only after he had received a clear signal that his counterparts had come to a decision that allowed further progress. Foxboro also sent proposals well ahead of time to allow the Chinese representatives ample time to develop their responses. The chief U.S. negotiator came to understand the importance of the representative of the central government (MMBI) at the negotiations and, on occasion, successfully appealed for help to the Center. When the joint venture's board was established, the first chairman was a ranking Shanghai government official, signifying the importance attached to Shanghai-Foxboro Company by the municipality.

Cummins Engine Company.   Cummins enjoyed an advantage in that a former vice president for research, an ethnic Chinese, was a graduate of Qinghua University, China's most prestigious technical institution, and was therefore very well connected with the engineering community. It was believed by Cummins personnel that visits and lobbying by this individual helped the company's cause. Nevertheless, conflicts and confrontations occurred during the long negotiation process, resulting in the disruption of contacts between the parties for months on end. The tensions between the parties may also have, in part, reflected the incomplete understanding, by the U.S. negotiators, of the Chinese side processes. More recently, Cummins has made significant progress in becoming more informed and sophisticated concerning the decision-making process in China. The Chinese American engineer who became a successful manager of the company's business activities in China and eventually in all of East Asia contributed significantly to this process.

Westinghouse.   Westinghouse negotiators expressed some uncertainty concerning the identity of many of the Chinese personnel who attended the meetings. Westinghouse personnel managed their time successfully by sending drafts of proposals well ahead of their arrival and thereby accommodated the Chinese side requirement of wide internal discussions prior to fruitful discussions and progress toward agreement. There was no indication of any success on Westinghouse's part to influence the Chinese decision-making process.

Combustion Engineering.   The lead negotiator for Combustion Engineering said that he had a feeling of isolation while conducting talks in China. He brought along a Chinese American engineer who had contributed to the company's original contacts with China and who served as language monitor and interpreter when needed. However, this individual refused to provide information on the atmosphere surrounding the meetings and would not make inferences on the cultural implications of what was said. As a result, the Combustion Engineering team worked under difficult conditions with limited information or insight as to how they were progressing. The chief negotiator thought, in retrospect, that he would have been better served had he accepted offers of representation from Chase Bank consultants who were fluent in Chinese. The U.S. personnel had some understanding of the composition of the Chinese negotiating team, but more complete insight only came later when the same individuals participated in the training program in the United States.

Duration of Negotiations

The time required to reach agreement represents an important criterion for successful management of the negotiating process since the cost to the company is certainly related to this factor. Foxboro initiated contacts in mid-1978, and all agreements were concluded by April 1982. However, in 1980 Chinese engineers were trained in the United States for technical support of Foxboro products imported to China. In December 1980 the most important agreement, the General Joint Venture Agreement was concluded. Cummins' negotiations lasted from August 1978 to January 1981. Westinghouse negotiations began in 1979, and agreement was concluded in September 1980, while Combustion Engineering took from February 1979 to November 1980 to reach agreement. Foxboro's total contacts and negotiations were the longest (almost four years) since it is significantly more difficult to conclude a joint venture agreement (Harrigan, 1985). The other three companies concluded their licensing agreements in about two years, give or take a few months.

STAGE THREE: START-UP

In technology transfer, expertise and knowledge are transmitted between individuals or organizations. The process includes the transfer of software as embodied in human expertise and in documentation, and the transfer of hardware such as materials, components, and equipment.

Combustion Engineering, as already discussed, co-designed the boilers to be built with the Chinese engineers. Thus the documentation required for manufacture was generated with the participation of the technology recipient or by its personnel alone under the supervision of the technology supplier. As a result, the entire complex of the usual problems associated with the transfer of paper in the other projects studied was minimized. The Chinese affiliate came to feel intrinsically connected with the transfer process from the beginning and did not feel threatened. The U.S. company was also quite free with documentation for technology not directly included in the project and allowed the Chinese trainees free access to it, permitting duplication and shipping to China without limitation. The procedure adopted was therefore free of many of the causes for conflict encountered by the other U.S. technology suppliers included in our study.

Foxboro managers also did not report any problem encountered in the course of the transfer of documentation and neither did the Chinese side to the project. Foxboro set up a special unit, termed the Documentation Control Center, for the sole purpose of tracking and maintaining the technical documentation transferred to the Shanghai joint venture. It may be that the establishment of the special unit helped prevent many of the difficulties encountered by other technology transfer projects.

Cummins Engine and Westinghouse encountered more difficulties. The requisite documentation was extensive and was scattered among different facilities requiring strenuous efforts to collect it and ship it to China. The process proved difficult to manage and delays occurred beyond the contractual schedules, causing stress and complaints from the Chinese side. However, the generated tension was well managed and complaints were addressed promptly. Chinese side sources claimed that Westinghouse failed to keep to agreed schedules due to domestic worker layoffs in response to lower order volumes. The technology recipient felt that the supplier had not been as responsive to its needs as it could have been.

Cummins ran into a special problem. The Chinese technology recipient complained of the absence of special sets of drawings (subassembly drawings), which, as it turned out, are required for manufacture in China but which are not used by Cummins. Tension was high but eventually the suspicion that the U.S. company had willfully withheld the drawings was allayed. However, the Chinese recipient of the technology had to generate the documentation before manufacture could be carried out. This is another example of fundamental differences between the technology cultures of the two countries, and clearer communication between the sides may have avoided the tension in the relationship created by the incident.

The documentation as received by the Chinese side from the U.S. technology supplier must usually undergo extensive modification before it can be used for the manufacturing process in China. Metrification is a trivial but labor-intensive example. In addition, paper has to be redrawn in accordance with a different convention governing projections and standards of materials. These factors were not an issue in the Combustion Engineering case, but they played a prominent part in the other three cases studied. Neither Cummins nor Westinghouse managers raised the issue during case study interviews, indicating possible lack of appreciation for its importance. The Chinese side stressed the burden of the work involved, and the researchers were told that Western European companies were considerably more attentive to the needs of Chinese affiliates in this regard than U.S. companies.

Training

Foxboro Company. At Foxboro, which established an equity joint venture, there was more freedom in designing training programs because the plant was not a union shop. In the first stage, 20 senior administrative and technical personnel were exposed to Foxboro's management philosophy and general orientation, followed by nine weeks of on-the-job training in each trainee's individual specialty. For example, the joint venture's designated chief accountant and his deputy received intensive training in the U.S. company's accounting practices, and they prepared a budget for the new enterprise's first year of operations. During the second stage, Chinese process engineers with prior experience in product lines similar to the ones to be transferred, received an introduction to Foxboro's design philosophy and then spent 12 weeks at the U.S. factory going through every step of the production process. They were asked to design a production process for the joint venture and to build a sample product using procedures suitable to the Chinese environment. To facilitate the design, the "Chinese environment," using a lower level of automation than prevalent at the U.S. plant, was simulated in a special area set aside for the training.

Training included the use of machines on the shop floor. Since Foxboro is not a union shop, no limitations applied to participation by outside visitors. At a later stage, Chinese engineers were trained in custom engineering (adapting instrumentation to the customer's needs) and design rationale, as required. Foxboro commissioned the writing of a glossary of technical terms to facilitate communications between technical personnel of the two affiliates. In its training program, the U.S. company tested the participants for understanding. The general approach adopted was to "train the trainers" such that the trainees were qualified to pass on the skills they had acquired in the United States to other technical personnel after their return to SFCL. In fact, incentives were instituted for passing on information and skills both in formal sessions and informally by evaluating the performance of work teams rather than of individuals. It can be stated that Foxboro's training methods were well suited to Chinese needs.

Cummins Engine Company. Cummins did most of their U.S. training in a special facility removed from the central Cummins plant. Cummins' work force is unionized, and therefore some limitations existed on activities of visitors in company plants. The U.S. company insisted on manufacturing training during the first stage with the aim to get production started as soon as possible. The Chinese side, on the other hand, wanted design training to take precedence, presumably to facilitate the preparation of production documentation to suit Chinese manufacturing requirements. A limited level of design rationale was included in later training programs, although there seems to have been some resistance on Cummins' part to the transfer of this type of information. Cummins mobilized as many Chinese American engineers as possible to lead training courses in the hope, largely fulfilled, that communications with the trainees would be eased. No testing of the trainees' assimilation of new knowledge or skills was carried out, and successful absorption of the presented material was largely the responsibility of the trainees.

Westinghouse. Westinghouse, as Cummins, convinced the technology recipient to agree to manufacturing training constituting the first stage of the training program, with design training being postponed to a later stage. A special problem was encountered in the Westinghouse licensing venture due to differences in the two sides' job definitions of design engineers. Chinese managers reported that in the United States such engineers carry out only the calculations required for the design but not the drafting. The Chinese design engineers, consequently, were not exposed to drawings and could not fully interpret blueprints after their return to China; they had to be sent back to the United States a second time to complete their training to the required level. In general, Westinghouse's use of classroom-type training was not perceived well by the Chinese side, who wanted on-the-job training.

Combustion Engineering. Combustion Engineering accepted the Chinese affiliates' preference for design training to come first. In this case, the Chinese engineers actually designed the boilers under U.S. personnel supervision at a U.S. facility, and they completed the detailed design work after returning to China supported by some consulting visits by U.S. engineers. It was the virtue of this strategy that enabled the Chinese personnel to prepare the manufacturing documentation to conform to Chinese specifications and standards. As a result, there was no need to invest the considerable time and energy necessary for adapting U.S.-prepared documentation, as was the case with the other projects studied. The strategy adopted by Combustion Engineering was suitable for the project, since it was necessary to custom design each boiler system to suit the local fuel supply characteristics.

General Training Problems. Training is the basis for one of the key success factors. Determining the success of training programs must be based on the efficiency with which skills are transferred to the Chinese trainees. The ability of trainees to perform the functions assigned to them once they return to their home enterprise and to train others provides a solid basis for assessing success, although the required information is not always available to the technology supplier. The principal tasks of the returning trainees include the development of the production process, installation of production equipment, maintenance, and custom engineering. Apart from training their own personnel, the trained engineers must also train the customers in applications and maintenance of the product. Another indicator is the level of effort needed for the training program and therefore its level of cost-effectiveness.

All U.S. technology suppliers alerted their trainers to use simple language in their efforts to bridge the language and culture gap. The Westinghouse affiliate sent interpreters to accompany the technical trainees, but this method proved to be inadequate since the interpreters' command of technical language and their technical background were too limited. Combustion Engineering teamed up Chinese engineers in pairs with an American trainer so that they could complement each other in English comprehension.

Logistical problems required appreciably more attention than had been the case for trainees from other countries. For example, none of the Chinese engineers knew how to drive, and therefore arrangements had to be made for their transportation. In Westinghouse's case, the Chinese officials in charge had to be convinced that bicycles were unsafe on U.S. highways, especially during rainy or winter weather. Cummins appointed a Chinese American to administer the program and to help the trainees adjust to the American environment. The U.S. companies had not foreseen these difficulties nor did they have sufficient appreciation of their scope, but in all cases they coped successfully with the problems as they arose. In all cases, the host companies were called upon to extend courtesies beyond the norm, and all of them provided the necessary services to help the Chinese visitors to overcome at least the initial difficulties of settling in, thus making important contributions to the training programs.

In all three licensing agreements studied, the U.S. technology suppliers accommodated their Chinese affiliates and agreed to very extensive training programs, even if they considered them excessive. For example, Westinghouse agreed to as many as 2,200 man-months of training for Chinese personnel in the United States, whereas by 1987, only 400 man-months were used. Cummins' affiliate requested and the U.S. company agreed to 600 man-months and after five years only 150 man-months had been utilized. The technology suppliers acted wisely in consenting to large volumes of training since the Chinese side has tended to view these training decisions as indicators of foreign companies' openness to transferring technology. For comparison it is interesting to note that Foxboro, in its joint venture, used about 250 man-months of training in the United States, supplemented by 400 man-months of on-site training in Shanghai. This comparison is significant since decisions to limit training programs were probably taken on the Chinese side in order to save foreign currency expenditure. In the Cummins case, for example, the cutbacks were considered excessive and were even believed to have caused delays in the technology absorption process by U.S. managers. However, in the Foxboro case, both sides were party to the decisions and therefore the extent of the training program may be accepted as balanced and appropriate to the requirements of the project.

An additional index for measuring the effectiveness of the training program in the United States is the amount of follow-up consulting necessary in China. However, it is difficult to ascertain how much of the consulting is required due to weaknesses in the training program. Such support is very expensive and, depending on the terms of the agreement, one of the primary stakeholders must bear the cost. Both Westinghouse and Combustion Engineering decided to provide consulting services at their own cost when the budget of the recipient had run out, even though it was the responsibility of the recipient to pay for such services. This was done because it was concluded that the additional services were required to ensure the success of the project and the U.S. companies felt that their reputations in China were at stake.

On-Site Consulting

A second success factor during this stage is the effective use of on-site consultants. This has been the cause of some friction because the parties view the function of a consultant differently. The Chinese often wish to limit the consultant to answering specific questions that they pose. On the other hand, the Americans tend to want to be fully informed on the process of absorption and of the problems encountered by the technology recipient, so that they can serve as support for project management. The degree to which agreement on the consulting function can be reached is a good indicator of successful cooperation between the parties to the technology transfer project. The subject is well worth serious attention by both sides, since consulting is likely to play a significant role over the lifetime of the technology transfer agreement.

Cummins engineers reported that their visits to the Chinese affiliate were too closely controlled in the beginning phase of start-up and were restricted to visitors' areas where the facilities available were not appropriate for technical discussion. Even the study of drawings was difficult since there were no tables on which to spread them. Also, technology recipient companies did not have enough funding to avail themselves of sufficient consulting to optimize the technology transfer process. As a result, the process was significantly delayed, in the opinion of American engineers having experience with similar projects in India. Westinghouse and Combustion Engineering felt that efficiency was sufficiently impaired by the Chinese affiliates' lack of funds, and so the former decided to provide consulting services at their own cost.

Foxboro had the advantage of a presence from the beginning in their joint venture plant in Shanghai and, as the figures cited above show, relied relatively heavily on on-site training, which we have called consulting in the context of the licensing projects.

Foxboro, Cummins, and Westinghouse executives have all noted that the transfer of management skills (software) has been difficult. Foxboro and Cummins further complained about the difficulty of transferring customer-oriented skills. The Chinese managers and other personnel did not have extensive background in these areas and successful training required persistent efforts. Much of this training was carried out at the facilities in China in order to reach as large a number of personnel as possible but it was limited in the licensing projects by limited availability of funding for U.S. personnel visits.

Transfer of Hardware

A third criterion during this stage concerns the transfer of hardware. The critical factor that can be measured is the time that elapses before actual production begins. Obviously, the evaluation of the quality of the product serves a complementary function. These last two criteria actually serve to judge the overall success of the venture, since production is, of course, a major aim of all concerned.

All projects successfully overcame the difficulties inherent in carrying the transfer process to production start-up. However, in the three licensing ventures the time taken was longer than judged appropriate in view of the moderate level of the technology and the good level of the technical personnel. Some of the causes have already been noted, and they were believed to be primarily managerial. For example, it took the recipients of the technology supplied by Cummins and Westinghouse about five years to reach a level of technical independence, as compared with three years it had taken for similar transfer in India.

Misunderstandings of Agreements

As an inverse indicator for successful transfer of technology, the sheer number of misunderstandings is critical. Misunderstandings have occurred for a number of reasons including differences in the degree of vertical integration in U.S. companies and in Chinese factories. The typical U.S. company has strength in a particular range of technologies and sources many components of its product outside the company. On the other hand, Chinese factories have been encouraged to build up self-reliance and independence from outside factors. As a result, the Chinese side has assumed that all technologies relevant to the product to be manufactured would be transferred. When the Chinese realized the limitations on the range of documented technologies, they were distressed, and strains developed between the partners.

In the Cummins case, the misunderstanding in this area was discovered before the agreement was signed, and the problem was successfully handled by referring the technology recipient to the U.S. company's suppliers. However, in the Westinghouse and Combustion Engineering cases, the Chinese factories and supervisory government agency only realized the limitations during the implementation phase when the documentation was transferred. The efforts required from both sides to overcome the resulting strains were correspondingly greater.

Coping with Renegotiation Demands

Renegotiation of contract terms during the implementation phase has been a subject of serious concern to U.S. companies and their successful handling therefore represents a fifth success criterion. Chinese authorities overseeing the implementation of individual companies have often required reconsideration of agreements in order to improve conditions for their side. The Chinese tend to regard the contract as considerably less final than does the typical U.S. executive.

In the case of Cummins, further financial concessions were required on Cummins' part in the face of pressure from the Chinese side. This episode was very upsetting to the U.S. company managers, who were unprepared for this pattern of business conduct. The relationship between the affiliates was saved by the ethnic Chinese engineer, who, by then, had become manager of the China Business Group, and who, through his insight into the opposite side's methods, encouraged his associates to deal objectively with the problem. Requests for renegotiations of nonfinancial conditions, such as modifications of the training programs or conditions governing component purchases, occurred in all cases studied, but agreement was reached successfully in spite of initial apprehension on the part of the U.S. companies. Foxboro and Westinghouse managers found that they had to resist Chinese tendencies to deviate from the agreement on an ongoing basis. At one point, the Shanghai-Foxboro Company Limited came under some pressure to accept newly assigned engineers, although there was no need for them and they had not been requested. The resident U.S. manager insisted on his rights to refuse to accept the engineers and found that this stance paid off in the long run in discouraging similar occurrences.

Westinghouse and Cummins have come under pressure by various Chinese individuals to broaden the scope of transfer of documentation and training to include technology not included in the agreement. Both companies resisted such pressures and kept to the limits specified in the agreement. Combustion Engineering, on the other hand, was quite free with the paper to which they allowed the Chinese trainees access, and thereby removed any cause for tensions arising from such demands. It is the policy of the company to allow relatively free transfer of its proprietary technologies among the broad family of licensees.

STAGE FOUR: MANAGEMENT OF THE ONGOING PROCESS

The fourth stage of the technology transfer process is referred to as management of the ongoing process. It carries the technology transfer to its final conclusion, which in the case studied here involved 15- to 20-year contracts. During this phase, the success indicators shift to long-term concerns.

Localization

During the start-up stage, all four cooperative agreements had some level of local content already in the initial product, namely, local assembly of imported kits. As the process of progressive localization proceeds during the latter phase of implementation, it extends into the fourth stage—management of the ongoing process. In the case of Westinghouse and Combustion Engineering, single large units were manufactured, and the level of local content for the initial product, called *verification units*, was decided upon during the negotiation stage;

implementation followed this plan. Nevertheless, localization for these projects is discussed together with Foxboro and Cummins in this section. Since most U.S. companies aim to develop local sources in China for materials and components in order to minimize foreign exchange expenditure, the extent of such localization constitutes a significant success criterion. Some companies have even encouraged their own suppliers to link up with potential component suppliers in China to further this aim. Often, the production process had to be modified to accommodate local conditions.

Localization of the production was successfully pursued, particularly by the Foxboro and Cummins affiliates, but the process required a great deal of attention and effort. The localization issue was, however, kept alive by the pressures of foreign exchange shortages felt by all stakeholders. Chinese companies are generally vertically integrated to a high degree, and therefore it was difficult to find specialized suppliers who were amenable to satisfying the needs. Locally produced parts and components also had to be "qualified" or tested for suitability as substitutes for imported items, and, as in the case of SFCL criteria of delivery, price and service were used in addition to the technical specifications. Cummins was successful in its demand that qualification should be subject to certification by its engineers and that performance tests had to be carried out with their participation.

Shanghai-Foxboro had to use its support structure in China to prevail on local factories to produce parts to Foxboro's stringent specifications, since the potential suppliers initially resisted producing special parts in small quantities as required. The joint venture expatriate engineers also advised Chinese suppliers on the production of these parts and thereby contributed to building the local infrastructure necessary for localized production by SFCL. Cummins similarly supported the development of the Chinese infrastructure. For example, the company trained personnel of the Chongqing Fuel Systems Plant as part of the regular licensing project training program, because this plant was a supplier of Chongqing Automotive Engines Plant, the primary technology recipient.

The degree of localization achieved varied. The first transferred product line of the Cummins affiliate reached a level of 30 percent localization in 1985, but during the following year (five years after start-up), a level of 80 percent was achieved. However, further increase was not expected in the near future. The verification units produced by the Westinghouse and Combustion Engineering affiliates were 30 and 70 percent localized, respectively, and further progress was expected with the manufacture of additional units. It appeared that Combustion Engineering was more attentive to the issue than Westinghouse. The former designed a transfer process that maximized localization and gave attention to modifications to further this aim. Combustion Engineering found that forgings in sizes required by its boiler production process were not available in China. It was therefore decided to use local castings and build the substitute parts by welding the castings.

Competition Between Technology Supplier and Recipient

Competition between the U.S. firm and its technology recipient (affiliate) is a general concern of technology suppliers. The successful technology transfer agreement must minimize the potential for such an adversary posture. Clearly, such occurrences are negative indicators, since they affect the supplier-recipient relationship adversely and may damage the supplier's competitive advantage. The rate at which the technology in question is expected to become outdated relative to the agreement period and the rate at which the Chinese affiliate may develop and become a producer of new technology are both indicators of the potential for such competition. At present, the technical work force resources of China are such as to expect a low level of competitive innovation for the near future, although their technology absorption capability is high (Schnepp). Foreign companies have also been concerned with the successful protection of proprietary technology and its restriction to the designated recipients. China enacted a patent law in 1985, but its effectiveness is yet to be tested. Loss of technology by leakage to third-party Chinese enterprises is therefore still a risk to be reckoned with and competition from companies outside the technology transfer project remains a possibility.

Cummins had the experience of competing for the same contract with its affiliate licensee in China. The problem was exacerbated by the fact that Cummins' bid was the lower one, which put in question the U.S. company's pricing of components sold to the licensee. At the time, the matter caused concern among Cummins personnel but the Chinese affiliate managers minimized the importance of competition between the affiliates when questioned by the researchers.

Establishing a Reputation as a Technology Supplier

One long-term indicator of success is establishing a reputation as a competent and reliable supplier of technology and high-quality products. Measurement of this indicator can be ultimately achieved by examining if the U.S. side succeeded in expanding its activity in China. We include in the factors considered the addition of further product lines to the original transfer project and the development of new technology transfer ventures.

Foxboro Company. In the beginning Foxboro manufactured simpler product lines and progressively introduced higher levels of sophistication. By the middle of 1984, customized systems technology, as opposed to single items, was transferred to SFCL, and in June 1985 the first two such sets were shipped to the end users. Managers held that the company's technology typically has a half-life of three to four years, and therefore quasi-continuous transfer of new technologies is necessary. However, the company did not plan to develop R&D activity in Shanghai. The plant's production volume and revenue have expanded greatly since start-up, but no plans for further ventures in China have been discussed in public.

Cummins Engine Company. Cummins transferred two product lines to its first licensee in Chongqing, but the agreement was limited to further innovations within these products. The company, however, first expanded its activity in China by establishing a service center, which was organized along lines similar but not identical to the Foxboro service company. In fact, Cummins planned to set up a number of such centers around the country and under the aegis of different ministries to overcome the barriers to horizontal communications among different administrative organizations. Cummins further concluded a licensing agreement with the No. 2 Automotive Works in Hubei Province, which has a production capacity roughly 10 to 15 times larger than the Chongqing affiliate. The company was also studying the feasibility of progressing to a joint venture with one of its Chinese affiliates.

Both Foxboro and Cummins are committed to contributing to the upgrading of local manufacturers in order to further localization as well as to develop their Chinese affiliates and possibly other local suppliers as low-cost sources of components. Both these aims promote foreign exchange balance and allow the expansion of the production volume, since it is expected that it will be necessary to continue using some U.S.-made components.

Combustion Engineering. Combustion Engineering has many years of experience as an international licensor and seems to be willing to accommodate foreign affiliates' methods of operation. For example, the company agreed to train the Chinese engineers in design at the beginning of the implementation phase as requested by the technology recipient and supervised the preparation of most of the documentation by the trainees themselves. Manufacture then began using the documentation so prepared; the Chinese side was eminently pleased by this mode of cooperation. Combustion Engineering personnel gave the impression that they were flexible and were at all times willing to consider the other side's opinions and priorities. The company seemed to be determined to overcome difficulties and to remain active in China.

Combustion Engineering had concluded eight technology transfer agreements by mid-1986 and was negotiating two more, two of these equity joint ventures. The Chinese companies came to realize that they required support in developing manufacturing capability in addition to the design capability they first sought to acquire. This recognition together with the successful affiliation with Combustion Engineering resulted in the expansion of the company's scope of activity in the country.

Westinghouse. Westinghouse did not establish further associations after recognizing that the Chinese consumer, the Ministry of Water Resources and Electrical Power (MWREP), at this time emphasized importing whole plants, whereas the U.S. company determined that this market was not profitable in the face of the international competition it faced in China. Westinghouse, by all accounts, had difficulties either in understanding the politics surrounding its technology transfer project or in adapting to China's policy changes. The managers connected with the project felt a sense of frustration; they felt stymied in their efforts to expand the company's involvement in power-generation equipment industry and decided to retreat in the face of barriers to further technology transfer to China in this sector.

Information on Chinese Market

Barriers to information flow are well recognized as an impediment to doing business in China. However, as the U.S. firm becomes more closely involved, the technology transfer project becomes an experimental tool for learning about the Chinese industrial and business environment as it pertains to the sector of interest. Further, the understanding of the decision-making process and its lead personalities can be advanced, and the technology supplier has improved opportunity to understand the Chinese enterprise's network of suppliers and end users of the product, as well as the lines of authority linking the technology recipient to local, regional, and central government agencies. These data help the Americans make a realistic assessment of the technology transfer process and of the opportunities for future expansion of the company's activities in China. Success can be measured by the adequacy of information available to the company's management for making policy decisions about current and future commitments to the Chinese market.

Foxboro Company. Foxboro through its resident expatriates (two at a time with three-year tours) has had ample opportunity to observe the China scene. However, it is known that the U.S. company prefers to operate through its Chinese partner as intermediary rather than testing its own accumulated knowledge of the environment. For example, at one point when several additional engineers had been assigned to the factory without such additional personnel having been requested, the expatriate plant manager requested the Chinese assistant manager to refuse to employ the new personnel. Nevertheless, the expatriates came to appreciate the difficulties encountered and concluded that the system and not the individuals set limits to what could be accomplished. Problems were encountered in procurement, and considerably greater effort was required to conclude sales and supply contracts than is needed in the United States. Also, the inadequacy of the telecommunications and transportation systems contributed substantially to the difficulties of doing business in China, the company learned.

Cummins Engine Company. Cummins clearly developed extensive data bases since this is a basic requirement for the successful business development achieved by this company in China. Cummins Engine Company has had the advantage of advice and guidance from a Chinese American engineer turned executive. He has established a network of personal and professional relationships and, by all indications, has been accepted by the Chinese managers and officials as sympathetic to their goals, even though he represents a foreign company. This is no small achievement since ethnic Chinese are often subject to opposing pressures and lose effectiveness as proponents of the U.S. side.

Combustion Engineering. Combustion Engineering, like Cummins, successfully expanded its activities in China and therefore is also assumed to have made considerable progress in understanding the business environment and infrastructure. The company also found that in more recent times Chinese personnel have become more communicative, thus considerably greater insight was gained into the problems encountered on the U.S. side. This progress was probably partly due to the growth of personal relationships, as hypothesized by U.S. personnel, but also China has become increasingly more open to revealing data on its economy, and particulars such as budgets are no longer the taboo subjects they once were.

Westinghouse. It is not clear to what extent Westinghouse built a satisfactory data base on China. If the company's decision to decelerate its activities in China in the area of power-generation equipment was based on sound information, the validity of this decision remains an empirical question that cannot be answered at this time. The judgment must be based on the success of other companies making profitable inroads into this market.

TECHNOLOGY TRANSFER ASSESSMENT

The effective transfer of the selected technology is the overriding indicator of success. Despite the fact that this criterion is one of the major overall indicators, it may not be possible to assess until the last phase of the technology transfer project. The criteria include product quality and the level of the recipient's technological independence. Other measurable indicators of success are the degree of localization of the product and the degree of independence of the technology recipient of the supplier company.

The technology transfer in all four cases studied were ultimately judged successful, but the technology absorption process took about 70 percent longer than anticipated (five years instead of three years) in the cases of the three licensing projects. By the middle of 1985, SFCL had manufactured and delivered to the end users both single-control instruments as well as customized control systems and, according to Foxboro's commitment, the quality was on a par with products of the parent company. Localization had also progressed significantly, although exact percentages were not given. Cummins' Chongqing affiliate in 1986 was producing diesel engines that were 80 percent localized and that were accepted by the U.S. engineers as of good quality. The Chinese affiliate was judged to have become technically independent of the technology supplier in the manufacture of the two product lines transferred. The Chongqing Plant received an award from the Chinese government for superior technology absorption achievement. The Westinghouse and Combustion Engineering affiliates completed their "verification units" in 1986–1987, and the U.S. companies were satisfied with the quality of the products. In these two cases, the products were large single units, comprising complete power stations in pairs, one of 300-megawatt capacity and the other of 600-megawatt. Because of the delays, the end user—Ministry of Water Resources and Electrical Power—decided to supplement local production with importation of complete plants from abroad, thereby diverting resources from the technology absorption projects.

The success in the transfer of management skills and the degree to which these skills have produced changes in the Chinese enterprise is a critical indicator of success. It has been common experience that it is harder to transfer management software than technical software and hardware. It is important to remember that the former skills are vital for the support of the technology transfer process. The management functions that must be in place to ensure efficient technology transfer include the basic organizational systems; procurement; inventory control; manufacturing planning and scheduling; accounting control; and, of particular importance, human resource management technology (Von Glinow and Teagarden, 1988).

Foxboro clearly had more control over the joint venture operation than did the other U.S. companies over their licensees' organizations. Foxboro introduced changes in SFCL's accounting practices and effected changes in the Chinese managers' attitudes. Previously there had been the tendency to interpret regulatory laws in the government's favor, whereas now they were asked to stress the company's advantage. The U.S. personnel found it difficult to bring about this reorientation. Management methods such as inventory control were also modernized, and a computerized management resource program was introduced. In the area of human resource management, basic changes were achieved at SFCL. The joint venture pioneered new policies in the handling of the payroll, recruiting and incentive programs, with workers being reviewed twice a year. These innovations have since been adopted by other joint ventures in China. In October 1985, at the time of the changing of the guard of the two Foxboro expatriate managers, emphasis was shifted from production orientation to marketing, indicating that the management of the U.S. parent company had decided that the technology absorption process was well advanced.

Westinghouse and Combustion Engineering personnel observed that their affiliates gained a great deal from training in customer service and scheduling, but much time had to be invested in this activity. The U.S. companies concluded that the Chinese personnel were capable, but the local infrastructure was limiting;

transportation was the principal difficulty and frustrated coordination of the delivery of materials. As a result, production efficiency suffered in spite of great efforts, which did, however, result in measurable improvements.

SUMMARY

In summary, the four U.S. firms' technology transfer projects should be judged moderately to highly successful. All four companies entered into technology transfer agreements over eight years ago, and they all have progressed through the difficult start-up phase of their transfers. Whether the long-term consequences are ultimately positive for these pioneers remains to be seen. The managers believe this to be so in three of the four cases. The Westinghouse case is distinct in that the company decided to retreat rather than to expand its activity in China in the power-generation sector, on the grounds of insufficient profitability.

CHAPTER 4

Implications for Management

As our cases reveal, the effective transfer of technology is a complex process during which expertise and knowledge are transferred between individuals and organizations. The process occurs at several different levels and includes the transfer of software, as embodied in human expertise and documentation, and the transfer of hardware, such as materials, components, and equipment. For the purpose of analysis, we have divided the technology transfer experience of a company into the sequential stages of development, negotiations, start-up, and management of the ongoing process.

LIMITING FACTORS

Due to the many facets of U.S.-China technology transfer, effectiveness demands that executives adapt to multiple factors including new communication styles, infrastructures, and decision-making philosophies. Crucial interactions occur among individuals on many different levels of intensity and intimacy. As we have noted in Chapter 3, *effective communication* is a basic necessity for the success of these interactions. However, language and cultural differences in behavior and customs often impede effective communication in the U.S.-China context.

In addition to communication difficulties, we have observed that there are two other underlying factors that govern U.S.-China interactions and limit the effectiveness of the process. One limiting factor, broadly termed *infrastruc-*

ture, covers the context within which Chinese enterprises operate. Infrastructure includes, for example, the high degree of vertical integration of Chinese enterprises; the vertical communication and control that exists in Chinese bureaucracy; the problematic communication and transportation system on which all enterprises must depend; and the availability, or lack thereof, of locally sourced, quality raw materials and components. One U.S. executive whom we interviewed in China aptly expressed this limitation as follows: "The biggest problems we face in China are not problems with our partner. They are problems that are bigger than both of us and that have to do with the surrounding infrastructure."

Another important limiting factor, which we call *decision-making process*, includes both the internal decision-making processes in Chinese enterprises, as well as the external controls that they are subject to from external organizations. Such government regulatory processes are particularly pronounced in issues relating to foreign exchange and technology export and import approvals. Even for issues as commonplace as travel abroad, Chinese enterprises are required to obtain approvals from relevant agencies before they can receive permission to send their managers or engineers for training to the United States. Since these permissions are obtained from agencies whose priorities may differ from those at the enterprise level, delays can frequently occur. It also means that the local enterprise does not have complete decision-making discretion on a particular issue.

These underlying factors, namely, communication, infrastructure, and decision-making processes, are manifested in different kinds of problems in the different stages of the technology transfer project. In the next section, we discuss the managerial implications of key problems in each technology transfer stage and suggest strategies for addressing them.

STAGE ONE: DEVELOPMENT

Background to Entry

The typical perspective that we have encountered in the U.S. executive contemplating entry into the Chinese market is that of China as the largest undeveloped and unexplored market. The Chinese market is also viewed as having tremendous growth potential. Most potential technology marketers are well aware of the restrictions on repatriation of profits in dollars, so most executives anticipate long-term profitability rather than short-term returns. Conventional market research data are not available, and extrapolations based on historical data are of limited use since the changes taking place in China in the 1980s and 1990s are unprecedented. As a result, most U.S. executives view China as a somewhat risky investment. However, the common conclusion seems to be that the risk of staying out of China is greater than the risk of entering with relatively limited information. To some extent, this conclusion is based on the assumption that the company that obtains an early foothold in the market of a centrally planned economy will have a first mover advantage and may be able to maintain a monopoly position with the support of the Chinese government. There also appears to be a desire to take advantage of the open door while it remains open.

The first of these presumptions, namely, the first mover advantage, needs to be viewed as being of somewhat limited benefit because the Chinese policy makers have demonstrated their reluctance to rely on a single company or, in fact, a single country. In automobiles, for example, projects now exist with American Motors of the United States, Peugeot of France, and Volkswagen of West Germany. It should be pointed out, however, that early entry still has an advantage because of the Chinese officials' emphasis on the history of a relationship and the related familiarity and trust that develop.

Chinese officials have demonstrated a tendency to approach the "best" company in terms of reputation of technological leadership. Foxboro, for example, had received consistently high ratings in technical and trade magazines and had a high proportion of R&D expense compared with total revenue. Chinese engineers have also shown a preference in going to the "source" of technology. This is demonstrated in the cases of Westinghouse, Cummins, Combustion Engineering, and Foxboro, all of which are the original patent holders. Since one of the high priorities of the Chinese officials is to acquire advanced technology, they also have a preference for companies with a demonstrated record of experience in international transfer of technology. The executives of another company approached by the Chinese side, Combustion Engineering, described their company as the largest licensing company in the world. Cummins too has licensees all over the world. Chinese engineers see established records of international technology transfer as an indicator that the company is willing to be open in sharing its technology with other companies and countries. Finally, historical contacts have proven to be very beneficial. Foxboro, for example, had done extensive business in pre-1949 China and had an established base of instrumentation products about which the older Chinese engineers were knowledgeable. Similarly, Cummins' engines were already in operation in China, and Chinese engineers were also in touch with Cummins' Japanese licensee, Komatsu, at the opening of negotiations with Cummins. During a period when numerous U.S. executives have attempted to initiate agreements and negotiations with Chinese officials, it is interesting to note the distinguishing characteristics of companies, as described above, with whom Chinese officials initiated contact.

Information on Chinese Market

Sources of information on the Chinese market are much more limited than those in other countries, although published official information and statistics have been on the increase in recent years.[1] Most foreign companies are still heavily dependent on their Chinese host organization for market information and for arranging visits to industrial plants. These visits are essential to determine which technology can be successfully absorbed and to gain an understanding of the level of professional expertise of available engineering and commercial personnel. These questions are still crucial, the relatively solid level of Chinese engineering notwithstanding. Recently, help has become available from Chinese consulting agencies, but the benefit derived from them is still not clear.

STAGE TWO: NEGOTIATIONS

Negotiation Imperatives

The typical profile of the U.S. company team during the negotiations is usually that of one key negotiator with considerable decision-making authority supported by one or two technical experts. Since the negotiation often follows a top management delegation to China, it attracts visibility and top management attention within the U.S. company. In addition, due to the prevailing language and cultural barriers, it gets, in the words of an experienced U.S. executive, "very lonely for the U.S. negotiator." For all these reasons, the dominant U.S. imperative is to "close the deal and sign a contract." As several experts have pointed out (Pye, 1982), however, an executive anxious to close the contract plays into the hands of the Chinese negotiators.

The typical Chinese imperative during the negotiations, by contrast, is to get to know and develop confidence in the potential U.S. partner. The Chinese negotiating team is usually a very large group by U.S. standards and includes representatives from different stakeholder groups. Some companies have encountered representatives on the Chinese negotiating team who do not utter a word during the entire negotiation process, but turn out later to be part of the engineering staff of the Chinese enterprise with which the U.S. company signs an agreement.

The Chinese decision-making process is very complex and requires multiple approvals and consensus from different stakeholders. Unlike the U.S. side, the Chinese negotiators are often conduits while the authority remains vested in the relevant ministry. Because of their inexperience in commercial negotiations, their somewhat turbulent history in acquiring technology from foreign countries, and their exploitation by foreign countries, a dominant driving force on the Chinese side is self-protection. This often means an inability or even unwillingness to take responsibility for individual decisions. The effective U.S. negotiator needs to understand this problem and needs to structure his or her negotiating

---

[1]The *China Yearbook* has been published annually since 1987; the *White Book on Science and Technology* was published for the first time in 1986.

strategy so as to help the Chinese counterparts address their concerns. Among other things, it means understanding the needs of the different stakeholders and "giving them enough so that they can report it with pride."

Executive Preparation

Many U.S. business-persons view foreign behavior patterns and foreign business practices as variations in American customs. To some extent, they are willing to learn about foreign social amenities and negotiating styles. However, the majority of U.S. negotiators in China have very limited knowledge when it comes to Chinese language and modes of communication, and far too many lack appreciation for the depth and antiquity of Chinese culture.

Many companies rely on local interpreters supplied by the Chinese host organization; others bring their own interpreters or monitors to ensure accuracy of translation. Whichever alternative is used, in most cases, the Chinese speaker is a Chinese American who has been on the staff of the company for some time. Such arrangements have not been uniformly successful. Although there have been examples of outstanding service by Chinese Americans, there is also evidence of friction between the Chinese American and the Chinese negotiators or even discomfort between the Americans and their colleague of Chinese origin.

Many misunderstandings have occurred because the U.S. negotiators lack familiarity with the Chinese mode of communication, which involves more ceremony and is much slower in coming to the point. In the early days after the opening up of China about 1978, outward appearances of Chinese personnel were particularly misleading. In their commercial dealings, Chinese feel that it is necessary to develop personal trust for their counterparts before they are willing to enter into a serious business relationship. They do not accept the practice of placing their trust solely in the legal force of the contract. Negotiations last longer, since this time period is used to promote interaction with the prospective affiliate, thereby developing familiarity and trust.

In order to minimize the difficulties encountered as a result of the barriers described above, U.S. personnel should ideally be trained to raise their awareness of, and sensitivity to, Chinese social customs and culture. Some knowledge of the Chinese language, even if rudimentary, is highly beneficial for promoting mutual confidence, beyond the obvious advantages derived from improved communication.

Consultation with the personnel of other companies who have China experience can also help. Such personnel regularly participate in public seminars where they present case study material and share their knowledge. Recently, a few consulting agencies have emerged in China, offering their services for market research, technology transfer guidance, general business consulting, and representation. The central problem in China is the lack of information concerning where to turn and whom to believe when claims are made concerning authority and expertise. A well-placed Chinese consulting agency can be of decisive value in this regard to a company seeking entry to China.

Negotiating Team

In the process of reaching agreement on a contract in China, it is best not to use lawyers until absolutely necessary. In most cases, Chinese personnel view U.S. lawyers as adversaries working according to standard rules and rigid formulas. Therefore, frustration and resentment on the Chinese side have resulted from the introduction of lawyers into the negotiating process. In addition, our experience suggests that it is desirable to nominate senior executives to the negotiating team. The advantages are twofold: (1) The Chinese are very conscious of hierarchy; a senior U.S. negotiator is more likely to be received by a Chinese counterpart who is also senior, which in turn, may expedite the Chinese decision-making process; and (2) a senior U.S. executive is more likely to have the authority and confidence to make decisions on the spot, if necessary. It is desirable, however, not to use the CEO as a negotiator. In the CEO's absence, the U.S. negotiator always has the backup excuse of needing to check back with the corporate office on a counter-offer as a way of buying breathing time.

As a final caution, it is advisable to have at least two core members on the U.S. negotiating team. The reasons are twofold: to counter (1) the psychological advantage of a much larger Chinese negotiating team, and (2) the emotions of isolation and loneliness experienced by a solo U.S. negotiator. The isolation is compounded by having to work in an alien language and culture for prolonged periods of time. The presence of even one other team member significantly helps counteract these effects.

Negotiating Contract Terms

Experience has shown that the technical terms surrounding a technology transfer project have been easier to negotiate than the commercial terms. We recommend leaving the negotiation of commercial terms to the last, after familiarity and mutual confidence between the parties have been developed. The primary reason for the disparity in the level of difficulty encountered during discussion of technical and commercial terms of an agreement lies in the nature of the Chinese personnel in these areas. Technical personnel are well trained, and, almost without exception, U.S. engineers found it easy to communicate with them once language problems have been overcome. On the other hand, Chinese commercial personnel commonly have gaps in their knowledge in general international practices, resulting in difficulties and delays.

Chinese officials appear uniformly resistant to the use of standard terms in contracts; they feel as though the terms are being dictated to them and that they have no control. Chinese personnel, regardless of their rank, insist on discussions and give and take. Chinese officials also insist on wholly owning a technology for which they have paid. As a result, they resist additional conditions that foreign providers of technology often wish to incorporate in transfer agreements. This resistance was codified in official regulations enacted in 1985.

Chinese personnel's limited experience in international commercial practices has been an important reason for difficulties encountered in the negotiation of commercial terms of agreements in China. While this is now changing, many Chinese negotiators have often wished to impress their superiors with having achieved a particularly advantageous agreement and therefore have felt constrained to be tough. They have, on occasion, frustrated U.S. negotiators by refraining from making a counter proposal, after asserting that the price demanded was too high. The foreigner negotiating financial terms in China must be prepared to be flexible and should assume that the Chinese side is well informed on competitors' prices on the world market.

The Chinese side obtains as much pricing information as possible, since it has relatively little sophistication in attaching commercial value to what is purchased. A U.S. company can help address the anxiety on the Chinese side by sharing some of its pricing information. One U.S. company, for example, broke through a stalemate on price negotiations by offering to show the Chinese team the prices it charges licensees in other parts of the world, thus proving that the Chinese were not being cheated.

Negotiating Process

Venue. The political and administrative structure prevalent in China predicates a decision-making process that is based on collective, not individual, responsibility. The Chinese team is always much larger than the foreign negotiating team, with many "work units" represented. Before progress can be made, the results of previous sessions must be extensively discussed with personnel not in attendance, and consensus must be reached. This is clearly an elaborate and time-consuming process which all but necessitates negotiations being conducted in China and which causes many of the delays that seem inexplicable to foreigners. Negotiating exclusively in China, however, gives the Chinese side the opportunity of playing host, and they appear to be astute at taking advantage of the situation by making the U.S. "guests" feel indebted and by setting the norms and agenda.

One of the Foxboro executives we interviewed made a case for balancing the negotiations venue between the United States and China. The advantages include an opportunity for the U.S. firm to showcase its facility, personnel, and products as a way of building credibility. In addition, it sends a signal that the U.S. company wants to "keep things mutual from the start." Resistance from the Chinese side to negotiations in the United States, however, can occur because of the foreign exchange expenditure involved. In addition, it may delay the decision-making process because the Chinese team frequently needs to obtain clarifications and approvals from the relevant ministries.

Decision processes. Although some changes may be possible on the Chinese side, a set of more realistic methods for dealing with local conditions can be proposed. Educating foreign negotiating personnel to fully understand China's political and organizational structure is essential for obtaining a full grasp of the decision-making process and for successfully coping with its manifestations. Many foreigners recommend submitting proposals in advance at every stage of the negotiations and waiting until the Chinese side is ready. When discussions falter and no progress seems forthcoming, U.S. negotiators often leave and propose renewal of the contact when either side has new ideas on how to proceed.

Recently, a U.S. company found it possible to affect the Chinese process by identifying a leading official and lobbying him directly. Chinese sources also advise that it is useful to approach a leading official directly. Such leaders are typically the rank of vice minister and do not usually participate directly in the negotiations. As a result, their identity is not known to the foreign team and must be discovered by diligent and persistent questioning during informal exchanges, best on a one-on-one basis.

U.S. executives are consistently advised to be patient in their negotiations with Chinese officials, because negotiations can last over two years. Based on our case studies, we add two caveats to the usual advice. First, it can be very helpful if the U.S. company maintains a flow of direct sales or other business in China, if at all possible, while the negotiations are in process. The ongoing revenue flow considerably reduces the pressure on the U.S. side to finalize a contract expeditiously and prevents the Chinese side from being able to use time to their advantage. Both Foxboro and Cummins used this tactic quite effectively; they continued to export products to Chinese customers throughout the negotiating period. Second, it is sometimes necessary to leave the negotiations. As we have noted in our case studies, sometimes the U.S. executive has lost patience and has told the Chinese that the U.S. company has already met all the feasible demands of the Chinese side and that if the terms are still not acceptable, he or she is ready to leave. Such an action can speed up negotiations. It almost appeared, from our cases, that the Chinese wanted to push the U.S. side to its limits without jeopardizing the agreement. Of course, for such a maneuver to be successful, the executive should have sufficient stature and authority, the company should possess a technology the Chinese side is eager to acquire, and the company should have already made a prolonged and genuine effort to accommodate Chinese needs.

Choice of Product and Technology

It is important that the transferred technology is "appropriate" for Chinese conditions. However, the choice often requires considerable negotiating before agreement is reached. A number of criteria of appropriateness have been identified.

The technology should be suitable for adaptation to local manufacture in order to allow localization of the product, that is, to make possible, ideally, the local production from locally sourced materials and components. In practice, localization up to 75 to 85 percent has been reached after five years of production in one case, but the results have varied from project to project. The lack of availability of manufacturing equipment identical to that used in the United States may be overcome by modification of the production process to accommodate local conditions.

The choice of technology to transfer has also been affected by the compatibility between the new technology and technology already in place in the Chinese plant. This criterion has influenced China's choice of the U.S. supplier. The level of R&D activity at the U.S. company is important, since most transfer agreements cover dynamic transfers in which the recipient plant receives information on new developments in the product line throughout the period of the agreement.

Marketability in the domestic Chinese market and in the international market are further conditions for success. The choice is complex since a certain technology may be important for China but may not be in demand in the world market. The Chinese side may ask for highly advanced technology in order to satisfy the government's demand for rapid modernization and also in order to produce an exportable product to generate foreign currency income. However, foreign companies transferring technology typically take the position that manufacturing should begin at a level of technology that is easily absorbed and that upgrading should be gradually accomplished, as technical and managerial skills improve. This argument has been difficult to sell Chinese negotiators, who may suspect the foreign company of withholding up-to-date technology.

In order to arrive at the best compromise between the various and often contradictory demands and preferences, it is essential that both sides are as fully aware of each other's capabilities as possible.

STAGE THREE: START-UP

Government Approvals

On both the Chinese and the American side, government approvals are required before implementation of an agreement can begin. In China, the corresponding industrial ministry must approve the agreement. This approval is usually implied when the contract is signed, since the ministry's representatives have usually been the principal negotiators. In addition, the Ministry of Foreign Economic Relations and Trade (MOFERT) must approve all aspects of the agreement, including and most important, the parts dealing with foreign exchange. However, Chinese officials point out that only one Chinese ministerial agency is required to approve the contract, whereas several departments are involved in the licensing process on the U.S. side. The Department of Commerce is the coordinating agency, but the Departments of Defense and State must also be consulted, and a consensus between all three must be reached. After all this has been completed, the license must also be approved by COCOM (Coordinating Committee), the Organization for Economic Cooperation and Development (OECD) coordinating agency located in Paris.

Training Issues

A number of levels of regulatory agencies are relevant to the training of Chinese personnel in the United States. On both sides, these agencies are external to the administrative structure of the participating organization. First, the training program content and a list of the trainees must be approved by the relevant Chinese ministry bureau before applications for passports can be submitted to the Public Security Bureau. Next, U.S. visas must be issued by the U.S. consulate. On the U.S. side, the training programs are subject to Department of Commerce licensing, as are all modes of technology transfer. The processes on both sides have often caused delays in the start-up of the U.S.-Chinese technology transfer projects, and serious investments of time and effort have been required to push them along.

Language and cultural barriers between training personnel and Chinese trainees have posed difficulties that have been overcome in various ways. Sometimes Chinese American engineers have been available to support the program. In other cases, interpreters have supplemented the language skills of the trainees. Some U.S. companies have used their facilities in Chinese-speaking areas for training (e.g., Singapore).

U.S. personnel engaged in technology transfer projects in China have also discussed their concern about the barriers to information flow within Chinese plants and between enterprises involved in the same or related industrial sectors. Defective information flow from trainees after their return to China to other personnel at the same plant has caused serious problems. Difficulties encountered because of barriers to information flow have been successfully combated by programming training schedules abroad and by structuring the information exchange after the trainees return to China. Incentives for information exchange have also been created by rewarding team performance rather than the skills of individuals.

Limitations on Component Technology

In China, the majority of factories have a high degree of vertical integration. U.S. companies, in contrast, have expertise in a core technology but source out to other companies the manufacture of many components. Because of their background, Chinese officials were often disappointed and angered when they discovered that a number of components were sourced from outside by the U.S. company and are therefore not included in the technology transfer agreement they have entered into. Sometimes similar considerations apply to the training of technical personnel. U.S. companies are constrained from implementing on-the-job training of production workers if their plant is a union shop; union rules prohibit access to machinery by unauthorized personnel.

Disputes or even demands for reopening negotiations have resulted from failure to communicate fully about exclusions of technology and training. Such crises can be prevented by exhaustive discussion on, and detailed listings of, the content of the projected technology transfer. To prevent friction caused by gaps in the technology transferred, related to components sourced from outside suppliers, the U.S. company can bring the Chinese side into contact with its outside sources and help achieve further agreements between these parties to supplement the original technology transfer process. It may also be possible to design complete transfer packages that include the suppliers and subcontractors of the principal foreign source of the technology.

Pricing of Components

The Chinese side has often been disturbed by the high cost of components charged by the foreign affiliate, whereas it judged the technology transfer fee to be reasonable. Many companies use this practice: charging low initial fees as an inducement and later compensating for the initial low return. This practice causes distrust to creep into the relationship between the two sides. Also, high payments that are long range exacerbate the persistent problem of shortage of foreign exchange on the Chinese side, and the high resulting cost makes marketing the product more difficult. The Chinese side is well advised to seek as much information as possible on costs from other affiliates of the foreign technology supplier and from worldwide sources. One U.S. executive commented that the Chinese officials do not understand the rationale behind pricing strategy. He used the analogy of an automobile that consists of 5,000 parts. The company may sell a small screw for $2 and a much larger part for $10. In his experience, the Chinese side would be angered by the relatively high price of the "small" screw. Such experiences emphasize the need for educating the Chinese counterparts on commercial logic and practices.

Documentation Transfer and Usability

Delays and defects in the documentation transfer process have occurred during start-up for a number of reasons. U.S. technology export licensing has often been delayed, preventing the U.S. firm from meeting the specified schedule. It has also been reported that U.S. companies have had difficulties collecting documentation, which is often located in a number of plants. These difficulties resulted in delays in the start-up of manufacturing and have increased costs.

In order to minimize the results of the delays caused by faulty documentation transfer, the parties to the cooperation should improve mutual understanding of each other's processes and problems. It is also beneficial to adapt the documentation management process of each side to promote closer linkage and thereby to increase the efficiency of the transfer. Management of documentation transfer is no simple task, since huge amounts ("car loads") of paper are involved. The problems encountered are exacerbated by Chinese demands to receive all documentation as soon as possible during the beginning of start-up, even though they are not able to absorb the material at the rate it is being transmitted.

The Chinese recipient of technology must overcome many difficulties before it can utilize the paper-clad technology transmitted by a U.S. supplier. To start with, all measurements must be translated into the metric system; this is routine but still time-consuming. Differences exist in the convention of projections used for technical drawings, necessitating the redrafting of thousands of technical plans. Also, some difficulties are encountered in translations of technical terms. Some companies have found it expedient to invest resources in the preparation of special guides and technical term glossaries in Chinese to minimize problems related to the language barrier.

The standards of construction materials (metals, plastics) in general use are quite different in the two countries. As a result, a number of tests and measurements must be performed before locally sourced materials can be substituted for the U.S. specified materials or before decisions can be made concerning which materials can be sourced in China and which must be imported.

The frequent modifications in design and the resulting near-continuous flow of new paper impose a major burden on many Chinese recipients under dynamic transfer agreements. This burden may be eased by updating at certain time intervals. At the end of each interval, a "package" of consolidated modifications is sent to the Chinese affiliate. Two-year intervals have been proposed as suitable.

Chinese Management Practices

From the point of view of Western observers and U.S. participants in U.S.-China technology transfer projects, the Chinese management system and its failures retard the technology absorption process. Chinese leaders have repeatedly stressed this problem and have acknowledged the need for management training. To illustrate the extent of the problem, one U.S. company in a joint venture reported that the transfer of inventory management and marketing and customer-orientation techniques was more difficult than the transfer of technical expertise. In fact, the prospect of obtaining training for Chinese personnel in foreign management techniques has been one of the primary attractions of joint ventures with foreign companies.

A number of measures can be recommended to advance the management capabilities of the Chinese personnel and organizations. Goal-setting and incentive systems should be instituted to motivate the Chinese personnel. Statistical quality control measures should be strengthened, using appropriate models.

Customer-oriented attitudes should be developed, and customer service systems must be organized. Inventory and distribution control must be improved. The transportation and coordination of supplies have been problem areas in Chinese organizations and have reduced production efficiency. Work-force utilization criteria must be developed, and utilization efficiency should be evaluated based on them.

STAGE FOUR: MANAGEMENT OF THE ONGOING PROCESS

Communication Among Affiliates

Communications and feedback problems have plagued many Chinese-foreign cooperative ventures, presumably because of cultural and language barriers. China suffers from a lack of tradition in openness and communication. As a result, the same difficulties also exist in communication among departments of the same enterprise and among different Chinese agencies and companies.

The tendency to limit communication has often restricted the amount of feedback from licensees to licensors when the latter did not have a presence in China. As a result, U.S. company personnel have reported that they lacked information on the progress of the technology transfer process, and they believed that absorption of the new technology suffered. In order to minimize the communications and feedback blockages, both sides to the cooperation must make determined efforts to raise the level of information exchange between the parties. In addition, it has been found that the presence of a U.S. expatriate manager can help overcome some of the communication barriers. However, it should be mentioned that expatriate personnel experience a special set of difficulties while living in a foreign environment whose culture and social structure are far removed from theirs. Some knowledge of the Chinese language, history, and literature is highly beneficial in bridging the wide cultural gap.

Dependence on External Authorities

Chinese agencies are organized along ministerial lines. Great barriers divide the ministries, severely restricting communication between agencies belonging to different ministries. In general, individuals make decisions dictated by the objectives of their own ministries or bureaus. Several kinds of problems arise from this organizational structure. For example, transportation schedules may be delayed by an official who follows his own instructions and is not responsive to the needs of the enterprise. Only intervention on the highest level may have an impact on the traditional ways of the Chinese bureaucracy.

Technical and production personnel are assigned rather than recruited by the employer, and the latter has only limited influence on this process, although the Joint Venture Law specifies the right of the Chinese-foreign cooperative enterprise to select and dismiss employees. It should be added, however, that foreign managers have said that they have been satisfied with the quality of engineering personnel. In addition, in recent times, it has been possible to screen personnel.

Foreign managers resident in China have said that it has been important for them to insist on their rights within their specific agreement or with reference to the Joint Venture Law. This is of particular importance when new regulations or agreements contravene traditional procedures.

In the final analysis, all business ventures in China, be they wholly Chinese-owned or be they partly or wholly foreign-owned, are limited in their operations by the ever-present specter of foreign exchange shortages. The growth potential and the ability to penetrate the local market are similarly constrained.

All foreign currency allocations in China must be approved by the relevant authority, even if such allocations are expressly stipulated in an agreement or contract concluded between a Chinese agency or enterprise and a foreign company. Repatriation of profits by the foreign partner to a joint venture or the purchase of parts are examples of transactions subject to this regulation. Some Chinese companies have been given preferential treatment in this regard in recognition of superior performance in the absorption of foreign technology.

The most direct strategy to alleviate the foreign exchange shortage is to engage in the export of the product for which technology was transferred. However, this is not always a viable alternative, either because of the lack of a market or because of the shortcomings in the product quality, particularly in the beginning stages of manufacture. Alternatively, the company may be able to engage in compensation trade or to set up trading packages in cooperation with other foreign enterprises. Recent Chinese government regulations allow an enterprise with foreign investment to charge Chinese customers in foreign currency for its products, subject to certain limitations. This possibility has been welcomed by joint venture companies. In addition, foreign exchange adjustment centers have been opened where it is possible to buy foreign exchange from other companies that generate foreign exchange, for example, tourist companies.

CONCLUSION

The essence of the insights discussed in this book have to do with the effective implementation of technology transfer across any two widely disparate cultures. Concerned with the issues that arise at different stages during the transfer of technology, we used U.S.-China technology transfer as the focal research to raise implications that are equally relevant to managing the technology transfer process with East European countries and LDCs. Problems of language, cultural differences, infrastructural constraints, different standards, and politicization of the technology transfer agreement complicate the transfer process in each of these settings. Overcoming these hurdles requires attention to the microdynamics of implementation, to the nuts and bolts of managing the transfer of expertise through people, documentation, and equipment, when confronted with cultural and other problems. Our exhaustive case studies enable us to address these issues in a very pragmatic way and provides lasting value that, we believe, transcends the Chinese political events of May 1989. Moreover, the key managerial lessons that are drawn from these case studies are relevant and applicable to all executive actions concerned with the effective internationalization of their firm's technology.

APPENDIX _____

Outline of Technology Transfer Case Studies

INTERVIEW PROTOCOL: KEY ISSUES

1. Development
   a. The position of China in the world strategy of the U.S. company
   b. Historical background of interaction with China and early contacts and approaches; research on China
   c. Information on Chinese industry and markets
   d. Political environment in China and in the United States and its changes; influence on U.S.-China cooperative ventures
   e. U.S. export licensing considerations
   f. Delegations to China and personnel involvement
2. Negotiations
   a. Place of negotiations and reasons for the choice
   b. Representative negotiating styles
   c. U.S. negotiating personnel; language and intercultural skills; interpreters and monitors of foreign language discussions
   d. Chinese negotiating personnel; their influence on style and rate of progress; relationship of Chinese negotiating team to production facility
   e. Choice of product and technology transferred; methods of technology transfer; choice of manufacturing site in China
   f. Management organization (in case of equity joint ventures)

g. Conclusion of agreements; content of agreements

h. Renegotiations

i. Payments to be made by each side; foreign exchange matters

3. Start-up

    a. Start-up personnel

    b. Technology transfer plan

        (1) Dissemination of documents

        (2) Training programs in United States and in China

        (3) Purchase of equipment for manufacturing process

        (4) Development of manufacturing process

        (5) Development of desired conditions in Chinese plants

        (6) Development of product

    c. Judgment of technology transfer progress and methods used for making the judgment

4. Management of the ongoing process

    a. Communications and feedback; U.S. company presence in China

    b. Manufacturing planning and schedules

    c. Performance—volume and quality

    d. Sourcing—raw materials and components (local in China or abroad, but not from U.S. partner or its affiliates)

    e. Customer service and customer training

    f. Competition between licensee and licensor's direct sales to China

    g. Servicing and customer follow-up of direct sales

    h. Availability and problems of foreign exchange; sources of foreign exchange

    i. Assessment of current position

    j. Development of technology transfer toward independent manufacture

5. Future Plans

    a. Further ventures of U.S. company in China

    b. Long-range strategy of U.S. company; evaluation of long-range prospects; description of long-range goals

    c. Long-range view of the China market and of the China business scene as seen by the U.S. company

REFERENCES

ADLER, NANCY J. 1986. *International Dimensions of Organizational Behavior*. Boston: Kent Publishing.

BALASUBRAMANYAM, V. 1973. *International Transfer of Technology to India*. New York: Praeger Scientific Press.

BARANSON, JACK. 1970. "Technology Transfer Through the International Firm," *American Economic Review, Papers and Proceedings*, May, pp. 435–40.

BARANSON, JACK, and A. HARRINGTON. 1977. *Industrial Transfers of Technology by U.S. Firms Under Licensing Arrangements: Policies, Practices and Conditioning Factors*. Washington, D.C.: Developing World Industry and Technology, Inc.

BEHRMAN, J., and H. WALLENDER. 1976. *Transfer of Manufacturing Technology Within Multinational Enterprises*. Cambridge, Mass.: Ballinger Publishing Company.

Beijing Review. 1989. February 6–12, p. 17. Reprinted in Foreign Broadcast Information Service *Daily Reports*, February 8, p. 16.

Beijing Television Service. 1989. Foreign Broadcast Information Service *Daily Reports*. January 13.

BELL, R. M., and S. C. HILL. 1978. "Research on Technology Transfer and Innovation." In *Transfer Processes in Technical Change*, eds. F. Bradbury, P. Jervis, R. Johnston, and A. Pearson, 225–74. Alphen aan den Rijn, The Netherlands: Sijthoff and Noordhoff.

BOYLE, K. A. 1986. "Technology Transfer Between Universities and the U.K. Offshore Industry," *IEEE Transactions on Engineering Management*, vol. EM–33 (no. 1), February, pp. 33–42.

BRADBURY, F. R., *et al.*, eds. 1978. *Transfer Processes in Technical Change*. Alphen aan den Rijn, The Netherlands: Sijthoof and Noordhoff.

CAMPBELL, NIGEL. 1986. *China Strategies*. Manchester, Great Britain: University of Manchester Press.

————. 1988. *China Strategies*, 2d ed. Manchester, Great Britain: University of Manchester Press.

China Business Review. 1989. Vol. 16 (no. 1), January–February.

China Economic Indicators (CEI) Database. 1988. November 10. Reprinted in Foreign Broadcast Information Service *Daily Reports,* November 10.

China Encyclopedia Yearbook, 1987. (Zhong'guo Baike Nianjian, 1987.) China Encyclopedia Publishing House, Beijing, pp. 310, 311.

CLOUGH, RALPH N. 1989. "Political Implications of Sino-Korean Trade." *China Business Review,* vol. 16 (no. 1), January–February, p. 42.

COHEN, JEROME A. 1988. *China Briefing, 1988.* Anthony J. Kane, ed. Boulder, Colo.: Westview Press, p. 107.

CONTRACTOR, FAROUK J., and T. SAGAFI-NEJAD. 1981. "International Technology Transfer: Major Issues and Policy Responses." *Journal of International Business Studies,* vol. 12 (no. 2), pp. 113–35.

DAHLMAN, CARL, and LARRY WESTPHAL. 1983. "The Transfer of Technology—Issues in the Acquisition of Technological Capability by Developing Countries." *Finance and Development,* December.

DAVIDSON, WILLIAM H. 1987. "Creating and Managing Joint Ventures in China." *California Management Review,* vol. 29 (no. 4), Summer, pp. 21–30.

DO ROSARIO, LOUISE. 1985. "Time to Pay the Piper." *Far Eastern Economic Review,* 27 August, p. 100.

DRISCOLL, R., and H. WALLENDER, 1981. "Control and Incentives for Technology Transfer: A Multinational Perspective." In *Controlling International Technology Transfer: Issues, Perspectives and Implications,* eds. R. W. Moxon and H. V. Perlmutter, 273–86. New York: Pergamon Press.

DUNNING, J. H. 1981. "Alternative Channels and Modes of International Resource Transmission." In *Controlling International Technology Transfer: Issues, Perspectives and Implications,* eds. R. W. Moxon and H. V. Perlmutter, 3–27. New York: Pergamon Press.

EVENSON, R. 1976. "International Transmission of Technology in Production of Sugar Cane." *Journal of Development Studies,* vol. 1, pp. 1–13.

FRAME, J. DAVIDSON. 1983. *International Business and Global Technology*. Lexington, Mass.: D. C. Heath.

FRANKENSTEIN, JOHN. 1988. *Current History,* vol. 87, p. 257.

FULK, JANET; EVERETT ROGERS; and MARY ANN VON GLINOW. 1988. "Managing Change Through Communication Technologies in Third World Countries." *Journal of Organizational Change Management,* vol. 1, no. 2.

Fund for Multinational Management Education. 1978. *Public Policy and Technology Transfer,* 4 volumes. Sponsored by FMME, Council of the Americas, U.S. Council of the International Chamber of Commerce, and George Washington University, March.

GEE, SHERMAN. 1981. *Technology Transfer, Innovation and International Competitiveness.* New York: John Wiley and Sons.

GODKIN, LYNN. 1988. "Problems and Practicalities of Technology Transfer: A Survey of the Literature." *International Journal of Technology Management,* vol. 3, no. 5.

GROW, ROY F. 1987. "American Firms and the Transfer of Technology to China: How Business People View the Process" (A Technical Report to the Office of Technology Assessment), February.

GROW, ROY F. 1988. "Acquiring Foreign Technology: What Makes the Transfer Process Work." In *Science and Technology in Post-Mao China.* Cambridge, Mass.: Harvard University Press.

GRUBER, WILLIAM H., and DONALD G. MARQUIS. 1969. *Factors in the Transfer of Technology.* Cambridge, Mass.: Massachusetts Institute of Technology.

HALL, G., and R. JOHNSON. 1970. "Transfer of United States Aerospace Technology to Japan." In *The Technology Factor in International Trade,* ed. R. Vernon, 305–58. New York: Columbia University Press.

HARRIGAN, KATHRYN R. 1985. *Strategies for Joint Ventures.* Lexington, Mass.: Lexington Books.

HENDRYX, S. R. 1986. "The China Trade: Making the Deal Work." *Harvard Business Review,* vol. 64 (no. 4), July–August, pp. 75–84.

HO, SAMUEL P., and RALPH W. HUENEMANN. 1984. *China's Open Door Policy.* Vancouver, B.C.: University of British Columbia Press.

HOFSTEDE, GEERT. 1980. *Culture's Consequences: International Differences in Work-Related Values.* Beverly Hills, Calif.: Sage.

IGNATIUS, ADI. 1989. *New York Times,* March 1.

JEREMY, D. J. 1981. *Transatlantic Industrial Revolution: The Diffusion of Textile Technologies Between Britain and America, 1790–1830s.* Cambridge, Mass.: Massachusetts Institute of Technology Press.

JOLLY, J. A., and J. W. CREIGHTON. 1975. *Technology Transfer and Utilization: A Longitudinal Study Using Benefit Analysis to Measure the Results from an R&D Laboratory.* Monterey, Calif.: Naval Postgraduate School.

KEDIA, BEN L., and RABI S. BHAGAT. 1989. "Cultural Constraints on Transfer of Technology Across Nations: Implications for Research in International and Comparative Management." *Academy of Management Review,* vol. 13 (no. 4), pp. 559–71.

KOIZUMI, T. 1982. "Absorption and Adaptation: Japanese Inventiveness in Technological Development." In *Managing Innovation: The Social Dimensions of Creativity, Invention and Technology,* eds. Sven Lungstedt and E. W. Colglazier, Jr., 190–206. New York: Pergamon Press.

LEE, DINAH. 1988. *Business Week,* August 8, p. 40.

LUBMAN, STANLEY B. 1986. "China's Economy Looks Towards the Year 2000" (U.S. Congress, Joint Economic Committee). Washington, D.C.: U.S. Government Printing Office, May 21.

MCINTYRE, JOHN R. 1986. "Introduction: Critical Perspectives on International Technology Transfer." In *The Political Economy of International Technology Transfer,* eds. John R. McIntyre and Daniel S. Papp, pp. 3–24. Westport, Conn.: Quorum Books.

MARLOW, P. 1985. Master's thesis. Bloomington, Ind.: Indiana University.

MARSHALL, JOHN. 1987. Presentation at the "Doing Business in China" Program, University of Southern California, May.

MARTON, K. 1986. *Multinationals, Technology and Industrialization.* Lexington, Mass.: D. C. Heath.

MASON, R. H. 1980. "A Comment of Professor Kojma's Japanese Type versus American Type of Technology Transfer." *Hototsubashi Journal of Economics,* January–February.

MAYER, C. S.; JING LUN HAN; and HUI FANG LIM. 1986. "An Evaluation of the Performance of the Joint Venture Companies in the People's Republic of China." Paper Presented at the University of Manchester, Manchester, England.

National Science Foundation. 1985. *Science Indicators.* Washington, D.C.: U.S. Government Printing Office.

New China News Agency 1986. January 24. Reprinted in Foreign Broadcast Information Service *Daily Reports,* January 27, p. K5.

_____. 1987a. January 31. Reproduced in Foreign Broadcast Information Service *Daily Reports,* February 2.

_____. 1987b. November 20. Translated in Foreign Broadcast Information Service *Daily Reports,* November 24, p. 29.

_____. 1988a. February 4. Translated in Foreign Broadcast Information Service *Daily Reports,* January 30.

_____. 1988b. April 28. Translated in Foreign Broadcast Information Service *Daily Reports,* April 29, p. 5.

_____. 1988c. November 1. Reprinted in Foreign Broadcast Information Service *Daily Reports,* November 2, p. 39.

_____. 1988d. November 4. Reprinted in Foreign Broadcast Information Service *Daily Reports,* November 7, p. 5.

_____. 1989a. January 13. Reprinted in Foreign Broadcast Information Service *Daily Reports,* January 17, p. 40.

_____. 1989b. February 4. Reprinted in Foreign Broadcast Information Service *Daily Reports,* February 7, p. 22.

NIEHOFF, ARTHUR H., and J. CHARNEL ANDERSON. 1964. "The Process of Cross-Cultural Innovation." *International Development Review, 6,* June.

NOVACK, DAVID E., and ROBERT LEKACHMAN, eds. 1964. *Development and Society.* New York: St. Martin's Press.

ORLEANS, L. A. 1987. *China Quarterly,* September, p. 444.

PUGEL, T. 1978. "International Technology Transfer and Neoclassical Trade Theory: A Survey" (Working paper). New York: New York University.

PYE, LUCIEN. 1982. *Chinese Commercial Negotiating Style.* Cambridge, Mass.: Oelgeschlager, Gunn and Hain.

_____. 1986. "The China Trade: Making the Deal." *Harvard Business Review,* vol. 64 (no. 4), July–August, pp. 76–80.

ROGERS, E. M., and F. F. SHOEMAKER. 1971. *Communication of Innovations.* New York: Free Press.

ROWELL, JOHN W. 1980. "The Way We Were, XII." *Power Team,* November–December, p. 7.

———. 1981. "The Way We Were, XIII." *Power Team,* January–February, p. 8.

SCHNEPP, OTTO. "Science and Technology Personnel in the PRC." Unpublished paper (see pg 36).

SIMON, DENIS F. 1987. "The Evolving Role of Foreign Investment and Technology Transfer in China's Modernization Program." In *China Briefing,* eds. J. S. Major and A. J. Kane. Boulder, Colo.: Westview Press.

State Science and Technology Commission. 1987. *China Science and Technology Policy Guide Book.* Science and Technology White Paper No. 1. Beijing, China: State Science and Technology Commission, Science and Technology Publishers, 1986. Translated in JPRS, China Science and Technology, issue 87–013, April 2.

STEWART, CHARLES T., and YASUMITSU NIHEI, 1987. *Technology Transfer and Human Factors.* Lexington, Mass.: Lexington Books.

SULLIVAN, ROGER. 1988. *Export Today,* vol. 4 (no. 2), March–April, p. 21.

TEECE, DAVID, ed. 1976. *The Multinational Corporation and the Resource Cost of International Technology Transfer.* Cambridge, Mass.: Ballinger Publishing Company.

United Nations Educational, Scientific, and Cultural Organization. 1986. *UNESCO Statistical Yearbook,* Table 3.10, pp. III–230.

United States–China Joint Sessions on Economic, Technological and Trade Development. 1988. Beijing: China, June. Many provinces made presentations for the benefit of visiting American businessmen in attendance.

United States Department of Commerce. 1988. Semi-annual *World Trade Outlook,* October.

VERNON, RAYMOND. 1966. "International Investment and International Trade in the Product Cycle." *Quarterly Journal of Economics,* May.

VON GLINOW, MARY ANN, and MARY B. TEAGARDEN. 1988. "The Transfer of Human Resource Management Technology in Sino-U.S. Cooperative Ventures: Problems and Solutions." *Human Resource Management Journal,* vol. 27 (no. 2), Summer.

VON GLINOW, MARY ANN; OTTO SCHNEPP; and ARVIND BHAMBRI. In press. "Assessing Success in U.S.-China Technology Transfer." In *Technology Transfer and International Business,* eds. T. Agmon and M. A. Von Glinow. New York: Oxford University Press.

Wall Street Journal. 1986. "Westinghouse Improves Profit Margins," January 31.

World Bank, 1986. *China, Management and Finance of Higher Education,* (Country Study). Washington, D.C.

World Bank, 1987. World Economic Report. Reprinted in MOR China Letter, 2, 11 (December 1988), p. 7.

YANG, LUJUN. 1989. *Shanghai Shijie Jingji Daobao (Economic Herald).* 1989. January 23. Translated in Foreign Broadcast Information Service *Daily Reports,* February 2, p. 52.

YOWELL, DIANE. 1988. *China Business Review,* vol. 15 (no. 5) p. 10.

YUN, KEN. 1989. "Crossing the Yellow Sea." *China Business Review,* vol. 16 (no. 1), January–February, p. 38.

Zhong'guo Baike Nianjian, 1987 (China Yearbook, 1987). 1988. Beijing, China: China Encyclopedia Publishing House.

Zhong'guo Tongji Nienjian, 1988 (China Statistical Yearbook, 1988). 1988. Ed. State Statistical Bureau. Beijing, China: China Statistical Publishing House.

ZHANG LANHUI, and JIYAN LUO. 1989. *Beijing Review,* February 6–12, p. 25.

DATE DUE

GAYLORD			PRINTED IN U.S.A.